Quantum Enigma

Physics Encounters Consciousness

Bruce Rosenblum and Fred Kuttner

OXFORD

UNIVERSITY PRESS

2006

OXFORD
UNIVERSITY PRESS

Oxford University Press, Inc., publishes works that further
Oxford University's objective of excellence
in research, scholarship, and education.

Oxford New York
Auckland Cape Town Dar es Salaam Hong Kong Karachi
Kuala Lumpur Madrid Melbourne Mexico City Nairobi
New Delhi Shanghai Taipei Toronto

With offices in
Argentina Austria Brazil Chile Czech Republic France Greece
Guatemala Hungary Italy Japan Poland Portugal Singapore
South Korea Switzerland Thailand Turkey Ukraine Vietnam

Published by Oxford University Press, Inc.
198 Madison Avenue, New York, New York 10016

www.oup.com

Oxford is a registered trademark of Oxford University Press

Library of Congress Cataloging-in-Publication Data
Rosenblum, Bruce.
Quantum enigma : physics encounters consciousness /
Bruce Rosenblum and Fred Kuttner.
   p. cm.
ISBN-13 978-0-19-517559-2
1. Quantum theory.   2. Science—Philosophy.
I. Kuttner, Fred.   II. Title.
QC174.13.R67 2006
530.12—dc22   2005029426

Wonderful Copenhagen, from the motion picture *Hans Christian Andersen*,
is used here as an epigraph to chapter 10.
Music and lyrics by Frank Loesser
© 1951, 1952 (Renewed) Frank Music Corp.
All Rights Reserved

9 8 7 6 5
Printed in the United States of America
on acid-free paper

We dedicate our book to the memory of John Bell, perhaps the leading quantum theorist of the latter half of the twentieth century. His writings, lectures, and personal conversations have inspired us.

Is it not good to know what follows from what, even if it is not necessary FAPP? [FAPP is Bell's suggested abbreviation of "for all practical purposes."] Suppose for example that quantum mechanics were found to resist precise formulation. Suppose that when formulation beyond FAPP is attempted, we find an unmovable finger obstinately pointing outside the subject, to the mind of the observer, to the Hindu scriptures, to God, or even only Gravitation? Would that not be very, very interesting?

—John Bell

Acknowledgments

During the preparation of our book we have greatly benefited from the suggestions, criticism, and corrections offered us by those who have read chapters as they were being prepared and revised. We gratefully acknowledge the help of Leonard Anderson, Phyllis Arozena, Donald Coyne, Reay Dick, Carlos Figueroa, Freda Hedges, Nick Herbert, Alex Moraru, Andrew Neher, and Topsy Smalley.

All along, our agent, Faith Hamlin, has given us crucial advice and warm encouragement. We very much appreciate her involvement in our book.

Contents

Quantum Enigma

1

Presenting the Enigma

Though what you're saying is correct, presenting this material to nonscientists is the intellectual equivalent of allowing children to play with loaded guns.

—A colleague's objection to our physics course, "The Quantum Enigma"

This is a controversial book. But nothing we say about *quantum mechanics* is controversial. The experimental results we report and our explanation of them with quantum theory are completely undisputed. It is the mystery these results imply *beyond* physics that is *hotly* disputed. For many physicists, this mystery, the quantum enigma, is best not talked about. It displays physics' encounter with consciousness. It's the skeleton in our closet.

One concern of physicists is that some people, seeing the solid science of physics linked with the mystery of the conscious mind, might become susceptible to all sorts of nonsense. We are sensitive to that problem and try to address it. Physicists can also be extremely uncomfortable with their discipline being involved with something so "unphysical."

Quantum theory is the most stunningly successful theory in all of science. Not a single one of its predictions has ever been wrong. Quantum mechanics has revolutionized our world. One-third of our economy depends on products based on it. However, this physics can look like mysticism. Quantum experiments display an enigma that challenges our classical worldview.

The worldview demanded by quantum theory is, to borrow the words of J. B. S. Haldane, not only queerer than we suppose, but queerer than we *can* suppose. Most of us share commonsense intuitions that deny the implications of quantum theory. For example, is it not just common sense that one object cannot be in two distant places at once? And, surely, what happens here is not affected by what happens at the same time someplace very far away. And does it not go without saying that there is a real world "out there," whether or not we look at it? Quantum mechanics challenges each of these intuitions by having (conscious) observation actually *create* the physical reality observed.

This idea is so hard to accept that some soften it by saying that observation *appears* to create the observed reality. Most physicists exploring the foundations of quantum mechanics today decline to sidestep the enigma with semantics and rather face up to what Nature seems to be telling us—though usually admitting that they don't fully understand it. After you see the archetypal quantum experiment you will be able to decide for yourself the extent to which the creation of reality by observation is just "apparent."

Since quantum theory works perfectly, for all *practical* purposes physicists can ignore—even deny—any mystery. But by doing so, we leave the aspects of the theory that most intrigue nonphysicists to misleading presentations such as, to take just a single example, the movie *What the Bleep?* (If you're unfamiliar with *Bleep*, see our comment early in chapter 14.) The *real* quantum enigma is not only more fascinating than the "philosophies" such treatments espouse, it is more bizarre. Understanding the true mystery requires a bit more mental effort, but it's well within the grasp of an intelligent nontechnical person.

The enigma we discuss is not just a way of looking at things, nor is it a new—or ancient—philosophical perspective. We describe straightforward physical phenomena that can be convincingly displayed to anyone. But with such demonstrations, we face an enigma that defies solution within our conventional worldview.

Though the quantum enigma has confronted physics for eight decades, it remains unresolved. It may well be that the particular expertise and talents of physicists do not uniquely qualify us for its comprehension. We physicists might therefore approach the problem with modesty—though we find that hard.

At the boundary where solid physics peters out, interpreting what's going on is controversial among physicists who think about it seriously. That physics has *encountered* consciousness cannot be denied. The continuing discussion by physicists of the connection of consciousness with quantum mechanics displays that encounter. Most interpretations of quantum theory show how the encounter with consciousness need not become a relationship. However, no interpretation evades it.

Here is how physics Nobel laureate Eugene Wigner once put it:

> When the province of physical theory was extended to encompass microscopic phenomena through the creation of quantum mechanics, the concept of consciousness came to the fore again. It was not possible to formulate the laws of quantum mechanics in a fully consistent way without reference to the consciousness.

Nevertheless, the physics community does not accept the study of consciousness itself as part of our discipline. And that is appropriate. Consciousness is too ill-defined, too emotion-laden. It is not the sort of thing we deal with in physics. But discussion relating quantum mechanics and consciousness will not go away.

In this book we describe the experimental facts and their accepted explanation by quantum theory. We then explore the resulting enigma and today's contending interpretations of what it all means. We do this as accurately as we can in nontechnical language. Fortunately, this is not too hard. The quantum enigma, conventionally called the "measurement problem," appears right up front in the simplest quantum experiment.

We have worked hard to make our book understandable. It is based on material we have developed over the past decade for presentation to nonscience students in what is now the most popular course in our physics department at the University of California, Santa Cruz.

Our bias, that the connection of the quantum enigma to the mystery of consciousness merits attention, will be obvious. Along with some of yesterday's and today's experts in quantum theory, we believe that quantum experiments hint at a worldview that has not yet been grasped. Astonishing insights might await us.

Only a minority of our physics colleagues shares our bias that the quantum enigma merits attention. Most physicists give the enigma little thought. Many are under the impression that it has been resolved by one or another of the "interpretations" of quantum theory. Most developers of those interpretations, however, still see a mystery.

A typical response, when a physicist is pressed to face up to the enigma, is that quantum mechanics simply shows that we must abandon naive realism. No one admits to being a *naive* realist. But if quantum theory denies the straightforward physical reality of atoms, it would also seem to deny the straightforward physical reality of chairs, which are *made* of atoms. Is Nature trying to tell us something? We strive to present honestly the facts and the ongoing controversy and come to the point where you can draw your own conclusions.

We often think in pictures. The many sketches in this book are mostly drawn by us. They're somewhat improved versions of what the two of us draw on our blackboards in explaining these slippery ideas.

How We Tell the Story

When I (Bruce) and another physics graduate student spent an evening with Albert Einstein, he tried to tell us of his misgivings with quantum theory. Unfortunately, our training was in the practical *use* of the theory and ignored its implications, which Einstein considered "spooky." Only decades later did I appreciate what we were unprepared to discuss that evening. That realization is part of the motivation for this book and is told of in our next chapter.

Our technology can demonstrate only with small things the weirdness of quantum mechanics. Therefore we illustrate a basic result of quantum mechanics with a parable, a fantasy. In it, a visitor to a land whose technology allows the display of quantum phenomena with large objects sees it demonstrated with large objects, with people.

Our intuitions about the way the world works are not all in our genes. They largely arose about three hundred years ago with the revolution in thinking that was started by Copernicus and Galileo and essentially completed by Newton. The worldview that is challenged by quantum mechanics is Newtonian, and the scientific attitude allowing that challenge is due to Galileo. In a single chapter we tell of the overthrow of Renaissance science by Galileo and the impact on our thinking of the Newtonian perspective, which is embodied in his universal law of motion.

Since we must speak about quantum phenomena in language common to both classical and quantum physics, we pick up in another chapter some basic ideas of electric fields, waves, and energy. We look at only enough of that classical physics to appreciate why physicists were *forced* to embrace quantum theory in spite of the weird things it says about the world.

The quantum first appears in Max Planck's explanations of the glow of hot bodies with his "desperate assumption" that violates the most basic notions of classical physics. Albert Einstein, taking Planck's assumption seriously, soon suggested that light is a stream of discrete particles. Since physicists could demonstrate the opposite, that light is a spread-out wave, Einstein's work was dismissed as "reckless." The plot thickened as this wave-particle duality was applied not just to light, but to *everything*.

Quantum theory emerged in its modern form in the 1920s with the Schrödinger equation as the *new* universal law of motion. The quantum enigma surfaced as the theory was seen to involve the *act of observation*—even *conscious* observation. Since this made quantum theory look like speculative philosophy, we emphasize the theory's down-to-earth practicality by seeing how one-third of our nation's economy depends on devices based on quantum mechanics.

In an imagined dialog, a physicist then displays the quantum enigma–physics' encounter with consciousness–to a "group of reasonable and open-minded people." They confront physics' "skeleton in the closet."

We now see how physics keeps its embarrassing skeleton in the closet with the "Copenhagen interpretation" of quantum mechanics. This interpretation builds the pragmatic case that since the theory *works*, everything is okay, for all practical purposes. It's the view we all accept in our practical research and teaching.

To explore the early unhappiness with Copenhagen, we tell Schrödinger's cat-in-the-box story and examine Einstein's profound argument that quantum theory presents an observation-created world only because the theory is incomplete. The only loophole in his argument was his out-of-hand dismissal of instantaneous connectivities that he called "spooky interactions." We then develop a nonmathematical version of Bell's theorem, which allows the demonstration that Einstein's spooky interactions indeed exist.

Other interpretations of quantum theory today compete with the Copenhagen interpretation—and with each other. It's a contentious field. But every interpretation encounters consciousness. With these interpretations of the *meaning* of quantum mechanics, we come to a boundary of physics as a discipline.

We now approach that boundary from the other direction. In today's surge of interest in consciousness, the issue of quantum mechanics arises in the "hard problem" of consciousness, the explanation of raw *experience*. We discuss two quantum theories of consciousness proposed by leading scientists. We then explore several joinings of the mystery of consciousness with the quantum enigma.

Our final chapter, "Consciousness and the Quantum Cosmos," takes the implications of the quantum theory to their almost impossible-to-believe logical conclusion. Wild speculations are inevitable, and you're invited to speculate for yourself.

2

Einstein Called It "Spooky"
And I Wish I Had Known

*I have thought a hundred times as much about the
quantum problem as I have about general relativity
theory.*

—Albert Einstein

*I cannot seriously believe in [quantum theory] because
. . . physics should represent a reality in time and
space, free from spooky actions at a distance.*

—Albert Einstein

In Princeton one Saturday in the 1950s, a friend asked his son-in-law and me
(Bruce) if we'd like to spend the evening with his friend Albert Einstein. Two
awed physics graduate students soon waited in Einstein's living room as he came
downstairs in slippers and sweatshirt. I remember tea and cookies but not how
the conversation started.

Soon Einstein asked about our quantum mechanics course. He was pleased
that we used David Bohm's text and asked how we liked Bohm's *philosophical*
treatment. We couldn't answer. We'd been told to skip that part of the book and
concentrate on the section titled, "The Mathematical Formulation of the Theory."
Einstein persisted, but the issues that concerned him were unfamiliar to us.
Our training was on the use of the theory, not its meaning. Our responses disappointed him, and that part of our conversation soon ended. It would be many
years before I understood Einstein's profound concern with the mysterious implications of the quantum theory, implications that he called "spooky" and that
he believed denied the obvious existence of the real world.

Quantum theory is not just one of many theories in physics. It is the frame-
work upon which all of today's physics is ultimately based. Einstein was bothered

9

by the theory's claim that if you observed an atom to be someplace, it was your *looking* that caused it to be there—it wasn't there before you saw it. Does that apply to big things? In principle, yes. Deriding quantum theory, Einstein once asked a fellow physicist, only *half*-jokingly, if he believed the moon is there only when you look at it.

Our book focuses on quantum theory's ostensible denial of a real world independent of its observation. But for years after that evening with Einstein I hardly thought about this weirdness, which physicists call the "measurement problem." As a graduate student I puzzled about the related "wave-particle duality." It's the paradox that, in one experiment, an atom could be *shown* to be a compact, concentrated thing; but with a different experiment, you *could have shown* that atom to be something spread out over a wide region. That seemed odd, but I assumed that if I spent an hour or so thinking it through, I'd see it all clearly. As a graduate student, I had more pressing things to do.

After a Ph.D. at Columbia and a postdoctoral year at Berkeley, I spent almost a decade as a researcher and research manager at a large electronics company. My work involved using quantum theory. But, like most physicists, I was no more concerned with what the theory *implied* than an automobile engine designer worries about the fact that the classical mechanics he uses is a deterministic theory that challenges free will.

Later, in the physics department at the University of California, Santa Cruz, graduate student Rob Shaw asked if we could put our experiments on quantum vortices in superconductors on hold for a couple of weeks so he could explore an idea in chaos theory. The couple of weeks turned into several years (and eventually a MacArthur "genius award" for Shaw).

Waiting to restart the superconductivity work, two students and I did calculations on how birds might sense Earth's magnetic field, a problem motivated by a biologist friend, who claimed his rats were affected by that magnetic field. I was skeptical: "You biologists don't understand how hard it is to detect a weak magnetic field." He came back with: "You physicists don't understand how complicated life can be." A telling point, not unrelated to the subject of this book.

I also started reading on the foundations of quantum theory and increasingly regretted being so unaware of the theory's strange implications that Einstein wanted to talk about at that meeting years ago. It was soon my turn to teach a physics course addressed to nonscience students. In such a course you can teach more or less what you wish, because you're not preparing students for the next physics course. I could focus on the quantum enigma, and I'd have an excuse to spend more time on my new interest.

I never went back to superconductivity. After a conference in Italy on the foundations of quantum mechanics, I was hooked on what I was unprepared to talk about that night in Princeton.

When I (Fred) encountered quantum mechanics in my junior year at MIT, I wrote Schrödinger's equation across the page of my notebook, excited to see the quantum equation that governed everything in the universe. Later that year, I tried to use quantum mechanics to analyze an experiment and was puzzled by the fact that an atom's north pole could point in more than one direction at the same time. Wrestling with this for a while, I gave up, figuring I'd understand it after I had learned more.

Coming to graduate school at the University of California, Santa Cruz, for my Ph.D., I met Bruce, who was then doing experimental work on superconductivity. He was friendly, but I wanted to do a theory dissertation, not lab work.

When the professor then teaching the graduate quantum mechanics course asked each student to write a paper on some aspect of the subject, I recalled an experiment stimulated by Bell's theorem that would display a strange, untested prediction of quantum theory. I found Bell's original paper in an obscure journal in the basement of the science library. Sometimes I understood what was said, and then a moment later I was confused. In the end, I just presented the mathematics.

For my Ph.D. dissertation I did a quantum mechanical analysis of phase transitions in magnetic systems. I had become facile in using quantum mechanics, but I no longer had time to think about what it meant. I was too busy trying to publish papers and get my degree.

After my Ph.D., I worked in Silicon Valley as a manager for a large electronics company and then for two start-ups. For years, I hardly thought about physics. Eventually, I returned to academia. By then, I had decided that only really fundamental studies interested me. Bruce and I soon recognized the mutual interest that became the focus of our research and led to this book.

When the two of us started to explore the boundary where physics meets speculative philosophy and where *some* claim it meets mysticism, colleagues were surprised. Our previous research areas were quite conventional, even practical.

The Skeleton in Physics' Closet

We started this chapter telling of Einstein's troubled concern with quantum theory. Let us now put that concern in perspective. Quantum theory was developed early in the twentieth century to explain the "mechanics"—the mechanism—governing the behavior of atoms. The energy of an atom was found to change only by a discrete quantity, a *quantum,* hence quantum mechanics, a term that includes both the actual experimental observations and the quantum *theory* explaining them.

Quantum theory is at the base of every natural science from chemistry to cosmology. We need it to understand why the sun shines, how TV sets produce pictures, why grass is green, and how the universe started in the Big Bang. Much of modern technology is based on devices designed with quantum mechanics.

Prequantum physics, classical mechanics, or classical physics, sometimes called Newtonian physics, is usually an excellent approximation for objects much larger than molecules, and it is simpler to use than quantum theory. But it is only an approximation, and it does not work at all for the atoms that everything is made of. Nevertheless, classical physics is basic to our conventional wisdom, our Newtonian worldview. But it's a worldview we now know is fundamentally flawed.

Since ancient times philosophers have come up with esoteric speculations on the nature of physical reality. But earlier generations had the logical option of rejecting such theorizing and holding to a straightforward, commonsense picture. Today, in light of facts demonstrated in quantum experiments, that commonsense view is no longer a logical option.

Can a worldview suggested by quantum mechanics have relevance beyond science? Consider a couple of other questions: Did Copernicus's denial that Earth was the center of the cosmos have relevance beyond science? What about Darwin's theory of evolution? The relevance of quantum mechanics is, in a sense, more immediate than either Copernican or Darwinian ideas, which deal with the long ago or far away. Quantum theory is about the here and now and even encounters the essence of our humanity, our consciousness.

Why then hasn't quantum theory had the intellectual and societal impact of those other insights? Perhaps because those others are easier to comprehend—and much easier to believe. You can roughly summarize the implications of Copernicus or Darwin in a few sentences. To the modern mind at least, those ideas seem reasonable. Try summarizing the implications of quantum theory, and what you get sounds mystical.

Let's try a rough summary anyway. To account for the demonstrated facts, quantum theory tells us that an observation of one object can instantaneously influence the behavior of another greatly distant object—*even if no physical force connects the two.* Einstein rejected such influences as "spooky interactions," but they have now been demonstrated to exist. Quantum theory also tells us that observing an object to be someplace *causes* it to be there. For example, according to quantum theory, an object can be in two, or many, places at once—even far distant places. Its existence at the particular place it happens to be found becomes an actuality only upon its (conscious) observation.

This seems to deny the existence of a physically real world independent of our observation of it. You can see why Einstein was troubled.

Erwin Schrödinger, a founder of modern quantum theory, told his now-famous cat story to illustrate that since the quantum theory applies to the large as well as the small, the theory is saying something absurd. Schrödinger's cat, according to quantum theory, could be simultaneously dead and alive—until your observation *causes* it to be either dead or alive. Moreover, finding the cat dead would create a history of its developing rigor mortis; finding it alive would create a history of its developing hunger—*backward in time.*

Anyone who takes the implications of quantum theory seriously would presumably agree that you can't accept it with equanimity. Niels Bohr, the theory's principal interpreter, tells us: "Anyone not shocked by quantum mechanics has not understood it." But a physicist setting out to design a laser or to explain the behavior of quarks, semiconductors, or stars must concentrate on his or her down-to-earth goal and ignore the theory's "shocking" implications. That is why, in teaching quantum mechanics to physics, chemistry, and engineering students, we avoid dealing with such things as the nature of reality or consciousness.

In fact, even mentioning such issues raises eyebrows. The story is told of a graduate student asking Richard Feynman: "Aside from being a tool for calculation, what actually *is* the quantum wavefunction?" The only response overheard was: "Shh! First close the door." As J. M. Jauch puts it: "For many thoughtful physicists, [the deeper meaning of quantum mechanics] has remained a kind of skeleton in the closet."

Back in the 1950s it was said that any nontenured faculty member in a physics department would endanger his or her career by showing interest in the implications of quantum theory. This is only somewhat less true today, but times are changing. Exploration of the fundamental issues in quantum mechanics increases today and extends beyond physics to psychology, philosophy, and artificial intelligence.

Because we focus on the "skeleton," some physics colleagues will disapprove of our book. But they will find nothing *scientifically* wrong with what we say. The physics facts we present are undisputed. Only when it comes to the meaning *behind* the facts is there argument. What those facts tell us about our world (and perhaps about ourselves) is today a contentious issue that extends beyond physics. There are intriguing hints of a connection of the world we call physical with that which we call mental.

The quantum enigma has challenged physicists for eight decades. Is it possible that crucial clues lie outside the expertise of physicists? Remarkably, the enigma can be presented essentially full-blown to nonscientists. Might someone unencumbered by years of training in the *use* of quantum theory have a new insight? After all, it was a child who pointed out that the emperor wore no clothes.

3

The Visit to Neg Ahne Poc
A Quantum Parable

If you're going to ham it up, go the whole hog.

—G. I. Gurdjieff

A few chapters will go by before we're ready to encounter the enigma posed by quantum mechanics. But we want to start out with a look at the paradox. Today's technology limits our displaying the quantum enigma to small objects only. But that is solely a technological limitation. Quantum mechanics applies to everything.

So we begin by telling a story in which a physicist visits Neg Ahne Poc, a land with a magical technology that allows demonstration of something like the quantum enigma, but with large objects—for example, a man and a woman—instead of atoms. Watch for what baffles this visitor to Neg Ahne Poc. His bafflement is the point of our parable. In later chapters you, too, will likely experience this bafflement that quantum mechanics invokes. In our story, the explanation that the Rhob offers to his visitor is essentially Bohr's Copenhagen interpretation of quantum mechanics.

Prologue by Our Self-Assured
Visitor to Neg Ahne Poc

Let me tell you why I'm slogging up this steep trail. Since quantum mechanics can make Nature appear almost mystical, some people become susceptible to wholly unjustified notions. They are led to accept supernatural foolishness.

A month ago I was with friends in California. People there seem particularly susceptible to such nonsense. My friends spoke of the *Rhob* in Neg Ahne Poc, a village high in the mountains. They claimed this shaman could display quantum-like phenomena with large objects. That is ridiculous, of course!

When I explained to them that such a demonstration is beyond even our advanced technology, they accused me of being closed-minded. I was challenged to investigate, and one of them, a dot-com billionaire, offered to fund my trip. Though my colleagues in the physics department urged me not to waste my time on a wild goose chase, I believe that a public-spirited scientist should expend some effort investigating unjustified notions to prevent their propagation. So here I am.

I'll look into this stuff with a completely open mind. I'll then debunk this nonsense when I get back to California. But while I'm in Neg Ahne Poc, I'll be discreet. This shaman's trickery is likely part of the local religion.

The trail becomes less steep and broadens to end suddenly in a spacious plaza. Our visitor has arrived in Neg Ahne Poc. He is relieved to see that his friend's long-distance arrangements have worked. His arrival is expected. He is warmly greeted by the Rhob and a small group of villagers.

Greetings, Curious Questioner, Careful Experimenter. You are a welcome visitor to our village.

Thank you, thank you very much. I appreciate the warm welcome.

We are happy to have you with us. I understand you come on a mission to seek truth. Since you are an American, I am sure you want it quickly. We will try to accommodate, but please sympathize with our unhurried ways.

Oh, I really appreciate that. I hope I will not be much trouble.

Not at all. I understand that you physicists recently—in the most recent century, as a matter of fact—have learned to demonstrate some of the deeper truths of our universe. But your technology limits you to working with small and simple objects. Our "technology," if you wish to call it that, can provide a demonstration with the most complex entities.

(ENTHUSIASTICALLY, BUT SUSPICIOUSLY) I'd be eager to see that.

I have made such arrangements. You will ask an appropriate question, and the answer to your question will then be revealed to you. I believe the procedure of posing a question and having an answer revealed is much like what you scientists call "doing an experiment." Do you wish this experience?

(LOOKS PUZZLED) Why, yes I do. . . .

I will prepare a situation to allow that experiment.

The Rhob motions toward two small huts about thirty yards apart. Outside each hut stands one of the Rhob's apprentices. Between the huts a young man and a young woman stand holding hands.

  Arranging our situation, "preparing the state" you would call it, must be done without observation. Please don this hood.

Our visitor places the soft black hood over his head and can now see nothing. After a few moments, the Rhob continues.

  The state is now prepared—please remove the hood. In one of these huts there is a couple, a man and a woman together. The other hut is empty. Your first "experiment" is to determine which hut holds the couple and which hut is empty. Do this by asking an appropriate question.

  Okay, in which hut is the couple, and which hut is empty?

  Very good, well done!

The Rhob signals his apprentice, who opens the door to the right hand hut to reveal a man and a woman arm in arm and smiling shyly. He subsequently has the door of the other hut opened showing it to be empty.

  Notice, my friend, you received an *appropriate* answer to your question. The couple was indeed in one of the huts. And the other hut was, of course, empty.

  (UNIMPRESSED, YET TRYING TO BE POLITE) Uh huh. Yes, I see.

  But I understand reproducibility is crucial to scientists. We will repeat the experiment.

Six more times this procedure is repeated for our visitor. Sometimes the couple is in the right-hand hut, sometimes in the left. Since our visitor is clearly getting bored, the Rhob stops the demonstrations and explains.

  (SOMEWHAT GLEEFULLY) Notice, my friend! Your asking the whereabouts of the couple caused the young man and young woman to be together, caused the couple to be concentrated in a single hut. If you have doubts, we could repeat this many more times.

  (ANNOYED BY HAVING TRAVELED SO FAR TO SEE AN APPARENTLY TRIVIAL DEMONSTRA-TION, OUR VISITOR IS FINDING IT HARD NOT TO OFFEND) My *questions* caused the couple to be in one hut or the other? Nonsense! Where you placed them while I was hooded did that. Oh, but, I apologize, that's beside the point. Thank you very much for your demonstration. I truly appreciate your trouble. But it's getting late; I must get down the mountain.

🧙 No, it is I who should apologize. I must remember: the attention span of Americans is short. I have heard that you actually choose the leaders of your nation on the basis of a number of thirty-second displays on a small glass wall.

Please, we now have a second experiment. You will ask a *different* question. You will ask a question causing the man and the woman to be in *separate* huts.

🚶 Well, yes, but I do have to be down . . .

Without waiting for our visitor to finish, the Rhob hands him the hood, and with a shrug our visitor dons it. After a minute or so the Rhob speaks.

🧙 Please remove the hood. Ask a new question to determine in which hut is the man and in which the woman.

🚶 Okay, okay, in which hut is the man and in which hut is the woman?

This time the Rhob signals his apprentices to open the huts at the same time. They reveal the man in the right hand hut and the woman in the left smiling at each other across the plaza.

🚶 V: (VISIBLY UNIMPRESSED) Uh huh.

🧙 Notice! You received an answer appropriate to the *new* question you asked, a result appropriate to the *different* experiment you did. The man was indeed in one hut and the woman was in another. Your question caused the couple to be distributed over both huts. We now display reproducibility by repeating this experiment.

🚶 Please, I *must* be leaving. May I just stipulate the reproducibility? (NOW WITH A SARCASTIC TONE OF VOICE) I concede that your "experiments" are all repeatable an arbitrarily large number of times with equally impressive results.

🧙 Oh, I *am* sorry.

🚶 (TAKEN ABACK BY HIS OWN DISCOURTESY) Oh, no, I apologize. I would be delighted to see a repeat of this experiment.

🧙 Well, maybe just two or three times?

And the demonstration is repeated three times.

🧙 R: You seem impatient. So maybe three times is enough to demonstrate that your asking the whereabouts of the man and the women separately caused the couple to be spread over both huts. Can you agree?

🕴 (BORED AND DISAPPOINTED, BUT SOMEWHAT SMUG) I surely agree that *you* can distribute the couple over the huts the way you wish. However, now I truly must be getting down the mountain. I had thought something else was to be demonstrated. But thank you very, very much for . . .

🕴 (INTERRUPTING) You have not yet seen the *final* version of these experiments. It is the *crucial* one that completes our demonstration. Let me do it for you—just twice. Only two times.

🕴 (CONDESCENDINGLY) Well, okay, two times.

Our visitor again dons the hood.

🕴 Please remove the hood and ask your question.

🕴 *Which* question should I ask?

🕴 Ah, my friend, you are now experienced with *both* questions. You may ask *either* of them. You may choose *either* experiment.

🕴 (WITHOUT MUCH THOUGHT) Okay, in which hut is the couple?

The Rhob has the door of the right hand hut opened to reveal a man and a woman hand in hand. He then has the door of the other hut opened showing it to be empty.

🕴 (A BIT PUZZLED, BUT NOT REALLY SURPRISED) Hmmm. . . .

🕴 Notice the question you asked, the experiment *you chose*, caused the couple to be in a single hut. Now let's try it again—for the second time—to which you did agree.

🕴 (QUITE WILLINGLY) Sure, let's try again.

Our visitor again dons the hood.

🕴 Please remove the hood and ask either question.

🕴 (WITH A TOUCH OF SKEPTICISM) Okay, this time I've decided to ask the *other* question: In which hut is the man, and in which hut is the woman?

The Rhob has his apprentices open both huts at the same time to reveal the man in the right hand hut and the woman in the left.

🕴 Hmmmmm. . . . (ASIDE—A SPOKEN THOUGHT) *Funny, he was able to answer my question this time too. Twice in a row. He could not know which one I would ask.*

🕴 Notice, my friend, whichever question *you choose* to ask is always appropriately answered. And now you wish to leave us.

🕴 Well, uh . . . , as a matter of fact, I would not mind at all trying this last experiment again.

🕴 Surely, I am delighted by your interest in the demonstration that no matter which experiment you choose, you get an appropriate result.

Our visitor once more dons the hood.

🕴 Please remove the hood and once again, ask either question.

🕴 Ok: This time, in which hut is the couple?

The Rhob has the door to the left hand hut opened to reveal the man and woman together. He then has the door of the other hut opened showing it to be empty.

🕴 You gave an appropriate answer to the question I chose three times in a row. Your luck is impressive!

🕴 It was not luck, my friend. The observation you freely choose will determine whether the couple will be together in one hut or divided in two.

🕴 (PUZZLED) How can that be? (EAGERLY, NOW) Can we try that again?

🕴 Surely, if you wish.

The demonstration is repeated, and our increasingly puzzled visitor requests yet further repetitions. Eight times he sees a result appropriate to the question he asked but inappropriate to the other question he could have asked.

🕴 (AN AGITATED ASIDE) *I can't believe this!* Please, I'd like to try this yet again!

🕴 I'm afraid it now *is* getting dark, and it is a steep climb down the mountain. Be assured that you will always get answers appropriate to the question you ask—appropriate to the situation your question caused to exist.

🕴 (MUMBLES AND LOOKS BOTHERED)

🕴 Something seems to trouble you, my friend.

🕴 How did you know which question I was going to ask when you placed your people in the huts?

🕴 I did *not* know. You could have asked either question.

🕴 (AGITATED) But, but . . . let's be *reasonable!* What if I had asked the question *not* corresponding to where the man and woman actually were?

🕴 My friend, did not your great Danish physicist, Bohr of Copenhagen, teach that science need not provide answers to experiments not actually performed?

Oh, yes, but come on—these people had to be either together or separated immediately *before* I asked my question.

I see what disturbs you. In spite of your training as a physicist, and your experience with quantum mechanics in the laboratory, you are still imbued with the notion that a physical reality exists independent of your conscious observation of it. Apparently even physicists find it hard to fully comprehend the great truth they have so recently gleaned.

Nothing you have seen violates your laws of quantum physics. Our "technology" has merely enabled us to extend to larger objects what your Bohr of Copenhagen has taught you to accept with equanimity for the small.

But good night, my friend. You have seen what you came to see. You must now leave us. Have a safe trip down the mountain.

(OBVIOUSLY BAFFLED AS HE TURNS TO LEAVE) Uh, yes, I will, uh, thank you very much, very much, I, uh, well . . . thank you . . .

(TALKING TO HIMSELF AS HE PICKS HIS WAY DOWN THE STEEP AND ROCKY TRAIL) Now let's see, there's *got* to be a reasonable explanation. If I asked where the *couple* was, he immediately showed me the couple to be in a single hut. But if I chose to ask where the man and woman each were *separately,* he immediately showed me one in each hut. Before I asked they had to be in one situation or the other. How did he do it?!

Was I tricked into asking the question that fit the setup he had arranged—like one of those forced-card tricks? No, I *know* my choices were freely made.

It's impossible! But I saw it. It's like a quantum experiment. Some have claimed that a conscious decision of what to observe creates the reality, but things such as consciousness shouldn't enter into physics. Anyway, quantum mechanics doesn't apply to big things such as people. Well, of course, that's not quite right. In *principle,* quantum physics applies to everything. But it's impossible to demonstrate such stuff with big things. Was I hallucinating?

How do I debunk this Rhob when I get back to California? And, oh my god! The guys back in the physics department will ask about my trip. *Ouch!*

There is, of course, no Neg Ahne Poc. What our visitor saw can't be demonstrated in the real world. But in later chapters we'll face a similar paradox: We will see that an object can be shown to be wholly in one place or, by a different choice of experiment, could have been shown to be distributed over two locations. Though present technology limits this display to small things, as technology advances it is being demonstrated for larger and larger objects. The orthodox view of the paradox, the Copenhagen interpretation of quantum mechanics, with Niels Bohr as its principal architect, is much like the view given by the Rhob in Neg Ahne Poc.

4

Our Newtonian Worldview
A Universal Law of Motion

Nature and Nature's laws lay hid in night:
God said, Let Newton be! And all was light.

—Alexander Pope

Quantum mechanics conflicts violently not only with our intuition but perhaps even with the scientific worldview we have held since the 1600s. Nevertheless, because quantum theory satisfies Galileo's criterion—that of experimental verification—physicists readily accept it as the underlying basis of all physics and thus of all science.

Galileo's bold stance created science, in any modern sense of that word. And within decades, Isaac Newton's discovery of a universal law of motion became the model for all rational explanation. Newton's physics led to the worldview that today shapes the thinking of each of us. Quantum mechanics both rests on that thinking and challenges it. To appreciate the challenge, we first take a look at the thinking.

Galileo insisted that scientific theories be accepted or rejected on the basis of experimental tests. Whether or not a theory fits with one's intuition should be irrelevant. This defied the scientific outlook of the Renaissance, which was, in fact, that of ancient Greece. Let's look at the problem Galileo faced in Renaissance Italy.

Greek Science: Its Contributions and Its Fatal Flaw

We owe the philosophers of ancient Greece credit for setting the scene for science by seeing Nature as explicable. When Aristotle's writings were rediscovered, they were seen as the wisdom of a "Golden Age."

Aristotle noted that everything that happens, even, say, the sprouting of acorns to become oak trees, is essentially the motion of matter. He therefore started by treating the motion of simple objects using a few fundamental principles. This is indeed the way we do physics today. But Aristotle's method for choosing fundamental principles made progress impossible. His mistake was assuming that they could be intuitively perceived as self-evident truths.

Here are a couple of them: A material object sought rest with respect to the cosmic center, which "clearly" was Earth. An object fell because of its desire for the cosmic center. A heavy object, with its greater desire, would therefore, *without doubt*, fall faster than a light object. In the perfect heavens, on the other hand, heavenly objects moved in that most perfect of figures, the circle. These circles were on the "heavenly spheres" centered on the cosmic center, Earth.

Greek science had a fatal flaw: *It had no mechanism to compel consensus.* The Greeks saw tests of scientific conclusions no more necessary than were tests of politics or aesthetics. Conflicting views could be argued indefinitely.

These thinkers of the Golden Age launched the scientific endeavor, but, without a method to establish agreement, progress was impossible. Though Aristotle established no consensus in his own day, in the late Middle Ages his views became the official dogma of the Church, mostly through the effort of Thomas Aquinas.

Aquinas fitted Aristotle's cosmology and physics together with the Church's moral and spiritual doctrine to create a compelling synthesis. Earth, where things fell, was also the realm of morally "fallen" man. The heavens, where things moved in perfect circles, were the realm of God and His angels. At the lowest point in the universe, at the center of Earth, was Hell. When, at the beginning of the Renaissance, Dante used this cosmological scheme in his *Divine Comedy*, it became a view that profoundly influenced Western thought.

Medieval and Renaissance Astronomy

From ancient times, the position of the stars in the sky foretold the change of the seasons. What, then, was the significance of the five bright objects that wandered through the starry background? An "obvious" conclusion was that the motion of these planets ("planet" means wanderer) foretold erratic human affairs and warranted serious attention. Astronomy's roots are in astrology.

In the second century A.D., Ptolemy of Alexandria described the heavenly motions so well that calendars and navigation based on his model worked beautifully. The astrologer's predictions—at least regarding the positions of the planets—were likewise accurate. Ptolemy's astronomy, with a stationary Earth as the cosmic center, required planets to move on "epicycles," complicated loopy curves made up of circles rolling on circles within yet further circles. King Alfonso X of Castile, having the Ptolemaic system explained to him, supposedly remarked: "If the Lord God Almighty had consulted me before embarking on Creation, I would have recommended something simpler." Nevertheless, the combination of Aristotle's physics and cosmology with Ptolemy's astronomy was accepted as both practical truth and religious doctrine, and enforced by the Holy Inquisition.

In the sixteenth century, an insight upsetting the whole apple cart appeared within the Church itself. The Polish cleric and astronomer Nicolas Copernicus felt Nature had to be simpler than Ptolemy's cosmology. He suggested that Earth and five other planets orbited a central, stationary sun. The back-and-forth wandering of the planets against the starry background was a result of our observation of them from an also-orbiting Earth. Earth was just the third planet from the sun. It was a simpler picture.

Simplicity was hardly a compelling argument. Earth "obviously" stood still. One *felt* no motion. A dropped stone would be left behind on a moving Earth! If Earth moved, since air occupied all space, a great wind would blow! Moreover, a moving Earth conflicted with the wisdom of the Golden Age. Such arguments were hard to refute. And, most disturbingly, the Copernican system was seen to contradict the Bible, and doubting the Bible threatened salvation.

Copernicus's work, published shortly after his death, included a foreword, probably added by a colleague, announcing his description as a mathematical convenience only—it did not describe *actual* motions. Any contradiction of the Church's teachings was disavowed.

A brilliant analysis some decades later by Johannes Kepler showed that accurate new data on the motion of the planets fit perfectly if he assumed that planets moved in elliptic orbits with the sun at one focus. He also discovered a simple rule giving the time it took each planet to orbit the sun. Kepler could not explain his rule, and he disliked these "imperfect" circles, but, rising above prejudice, he accepted what he saw.

Kepler did great astronomy, but science did not guide his contemporary worldview. Initially, he considered the planets to be pushed along their orbits by angels, and as a sideline he drew horoscopes, in which he likely believed. He also had to take time from his astronomy to defend his mother from accusations of witchcraft.

Galileo's New Ideas on Motion

In 1591, at only twenty-seven years of age, Galileo became a professor at the University in Padua, but he soon left for a post at Florence. Today's university faculty would understand why: He was offered more time for research and less

Figure 4.1 Galileo Galilei. Courtesy Cambridge University Press

teaching. His talents included music and art as well as science. Brilliant, witty, and charming, Galileo could also be arrogant, brash, and petty. We could envy his skill with words. He liked women, and they liked him.

Galileo was a convinced Copernican. That simpler system made sense to him. But unlike Copernicus, Galileo did not merely claim a new technique for calculation; he argued for a new worldview. A humble approach was not his style.

The Church had to stop Galileo's call for independent thought—the business of the Church was saving souls, not scientific validity. Found guilty by the Holy Inquisition and given a tour of the torture chambers, Galileo recanted his heresy of a sun-orbiting Earth. For his last years, Galileo lived under house arrest—a lesser penalty than that of the Copernican Giordano Bruno, who was burned at the stake.

Recantation notwithstanding, Galileo knew that Earth moved and that Aristotle's explanation of motion could not survive on a moving Earth. Friction, not desire for rest in the cosmic center, caused a sliding block to stop, and air resistance, not less desire, caused a feather to fall more slowly than a stone.

Contradicting Aristotle's claims, Galileo asserted: "In the absence of friction or other impressed force, an object will continue to move horizontally at a constant rate." And: "In the absence of air resistance heavy objects and light objects will fall at the same rate."

Galileo's ideas were obvious—*to him*. How could he convince others? Rejecting Aristotle's teaching for the motion of matter was not a minor issue. Aristotle's philosophy was an all-encompassing worldview. Reject a part, and you appear to reject it all.

The Experimental Method

To compel agreement with his ideas, Galileo needed examples that conflicted with Aristotle's mechanics—but that conformed to his own ideas. But looking around, he could see few such examples. His solution: *create them!*

Galileo would contrive special situations: "experiments." An experiment tests a theoretical prediction. This may seem an obvious approach, but in that day it was an original and profound idea.

In his most famous experiment, Galileo supposedly dropped a ball of lead and a ball of wood from the leaning Tower of Pisa. The simultaneous click of the wood and the thud of the lead proved the light wood fell as fast as the heavy lead. Such demonstrations gave reason enough, he argued, to abandon Aristotle's theory and to accept his own.

Some faulted Galileo's experimental method. Though the displayed facts could not be denied, Galileo's demonstrations were contrived situations, therefore insignificant because they conflicted with matter's intuitively obvious nature. Moreover, Galileo's ideas *had* to be wrong because they conflicted with Aristotelian philosophy.

Galileo had a far-reaching answer: Science should deal only with those matters that can be demonstrated. Intuition and authority have no standing in science. *The only criterion for judgment in science is experimental demonstration.*

Within a few decades, Galileo's approach was accepted with a vengeance. Science progressed with vigor never before seen.

Reliable Science

Let's agree on some rules of evidence for accepting a theory as reliable science. They will stand us in good stead when we consider quantum theory and will serve as a test for any ideas that theory might inspire.

But first, a remark on the word "theory": We speak of quantum *theory* but of

Newton's *laws*. "Theory" is the modern word. We can't think of a single twentieth- or twenty-first-century "law" in physics. Though "theory" is, at times, used in science for a speculative idea, it does not necessarily imply uncertainty. Quantum *theory* is, as far as is known, completely correct. Newton's *laws* are an approximation.

For a theory to compel consensus, it must, first of all, make predictions that are testable with results that can be displayed objectively. It must stand with a chip on its shoulder challenging would-be refuters.

"If you're good, you'll go to Heaven." That prediction may well be correct, but it is not objectively testable. Religions, political stances, or philosophies in general are not scientific theories. Aristotle's testable theory of falling—that a two-pound stone will fall twice as fast as a one-pound stone—is a scientific theory, albeit a wrong one.

A theory making testable predictions is a candidate for being reliable science. Its predictions must be tested by experiments that challenge the theory by attempting to refute it. And the experiments must be convincing even to skeptics. For example, theories suggesting the existence of extrasensory perception (ESP) make predictions, but so far, tests have not been convincing to skeptics.

To qualify as reliable science, a theory must have many of its predictions confirmed without a single disconfirmation. A single incorrect prediction forces a theory's modification or abandonment. This scientific method is hard on theories—one strike and you're out! Actually, no scientific theory is ever *totally* reliable—it is always possible that it will fail some future test. A scientific theory is, at best, *tentatively* reliable.

The scientific method, setting high standards for experimental verification, is hard on theories. But it can also be hard on us. If a theory meets these high standards, we are obligated to accept it as reliable science—no matter how violently it conflicts with our intuitions. Quantum theory is our case in point here.

The Newtonian Worldview

Isaac Newton was born in 1642, the year Galileo died. With the wide acceptance of the experimental method, there was a sense of scientific progress, though Aristotle's erroneous physics was still often taught. The Royal Society of London, today a major scientific organization, was founded in 1660. Its motto, *Nullis in verba*, translates loosely as, "Take nobody's word for it." It would have delighted Galileo.

Newton, a handy fellow, was supposed to take over the family farm. But more interested in books than plows, he managed to go to Cambridge University by working at menial tasks to help pay his way. He did not shine as a student,

Figure 4.2 Isaac Newton. Courtesy Cambridge University Press

but science fascinated him—"natural philosophy," it was then called. When the Great Plague forced the university to close, Newton returned to the farm for a year and a half.

Young Newton understood Galileo's teaching that on a perfectly smooth horizontal surface a block, once moving, would slide forever. A force is needed only to overcome friction. With a greater force, the block would speed up; it would accelerate. Galileo, however, accepted the Aristotelian concept that falling was "natural" and needed no force. He also had planets moving "naturally" in circles without any force. Galileo just ignored the ellipses discovered by his contemporary, Kepler. To conceive his universal laws of motion and gravity, Newton had to move beyond Galileo's acceptance of Aristotelian "naturalness."

Newton tells that his inspiration came as he watched an apple fall. He likely asked himself: Since a force was needed for *horizontal* acceleration, why not a force for *vertical* acceleration? And if there's a downward force on an apple, why not on the moon? If so, why doesn't the moon fall to Earth like the apple?

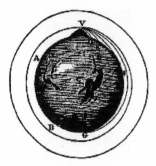

Figure 4.3 Newton's drawing of a cannon on a mountain

In Newton's famous cannon-on-a-mountain picture, the dropped cannonball falls straight downward, while those fired with larger velocities land farther away. If a ball is fired fast enough, it will miss the planet. It nevertheless continues to "fall." It continues to accelerate toward Earth's center while it also moves "horizontally." It thus orbits Earth. As the cannon ball comes around, the cannoneer had better duck!

The moon doesn't crash to Earth only because it, like that fast cannon ball, has a velocity perpendicular to Earth's radius. Newton realized what no one had before: The moon *is* falling.

The Universal Law of Motion and, *Simultaneously,* a Force of Gravity

Galileo thought that uniform motion without force applied only to motion that was parallel to the surface of Earth, in a circle about Earth's center. Newton corrected this to say that a force is needed to make a body deviate from a constant speed in a *straight line.*

How much force is needed? The more massive the body, the more force should be needed to accelerate it. Newton speculated that the force needed was just the mass of the body times the acceleration produced, or $F = Ma$. It's Newton's universal law of motion.

In Newton's day, however, there seemed to be a counterexample: Falling was a downward acceleration, apparently without an impressed force. Young Newton had to simultaneously conceive two profound ideas: his law of motion and the force of gravity.

When the plague subsided, Newton returned to Cambridge. Isaac Barrow, then Lucasian Professor of Mathematics, was soon so impressed with his one-time student that he resigned to allow Newton to take the Lucasian chair. The quiet boy became a reclusive bachelor. (Celibacy was required of Cambridge faculty—no longer so.) Newton was reserved and moody and was often angered by well-intended criticism. You'd rather spend an evening with Galileo.

Newton's ideas needed testing. However, his force of gravity between objects that he could move about on Earth was far too small for him to measure. So he looked to the heavens. Using his equation of motion and his law of gravity, he derived a simple formula. A chill no doubt ran down his spine when he saw it—his formula was precisely the unexplained rule Kepler had noted decades earlier for the time it took each planet to orbit the sun.

Newton could also calculate that the orbital period of the moon was consistent with a falling object gaining a speed of ten meters per second each second—something experimentally shown by Galileo. His equations of motion and gravity governed apples as well as the moon—on Earth as it is in the heavens. Newton's equations were universal.

Principia

Newton realized the significance of his discoveries, but controversy over the first paper he ever wrote had seriously upset him. The idea of publishing now terrified him.

Some twenty years after his insights back on the farm, Newton was visited by the young astronomer Edmund Halley. Knowing others were speculating on a law of gravity that would yield Kepler's elliptical orbits for the planets, Halley asked Newton what orbits his law of gravity would predict. Newton immediately answered, "ellipses." Impressed by the quick response, Halley asked to see the calculations. Newton could not find his notes. "While others were still seeking a law of gravity, Newton had already lost it."

After Halley warned him that others might scoop him, Newton spent a furious eighteen months producing *Philosophiae Naturalis Principia Mathematica*. What is now just referred to as *Principia* was published in 1687 with Halley footing the bill. Newton's fears of criticism were realized; some even claimed he stole their work.

Though *Principia* was widely recognized as the profound revelation of Nature's laws, being mathematically rigorous and in Latin, it was little read. But popularized versions soon appeared. *Newtonianism for Ladies* was a best-seller. Voltaire, aided by his more scientifically talented companion, Madame du Châtelet, in his *Elements of Newton* claimed to "reduce this giant to the measure of the nincompoops who are my colleagues."

The revealed rationality of Nature was revolutionary. It seemed to imply, in principle at least, that the world should be as understandable as the mechanism of clocks. This was later dramatically demonstrated by Halley's accurate prediction of the return of a comet. Until then, comets were commonly thought to foretell the deaths of kings.

Principia ignited the intellectual movement known as the Enlightenment. Society would no longer look to the Golden Age of Greece for wisdom. Alexander Pope captured the mood: "Nature and Nature's laws lay hid in night: / God said, Let Newton be! And all was light."

When he needed better mathematics, Newton invented calculus. His studies of light transformed the field of optics. He held the chair in Parliament then reserved for Cambridge. He became Director of the Mint and took the position seriously. In his later years, Sir Isaac—the first scientist ever knighted—was perhaps the most respected person in the Western world. Paradoxically, Newton was also a mystic, immersing himself in supernatural alchemy and interpretations of Biblical prophecies.

Newton's Legacy

The most immediate impact of the Newtonian worldview was the breakup of the late medieval synthesis of the physical and the spiritual. While Copernicus had, unintentionally perhaps, initiated the destruction of this Church-sponsored relationship by denying Earth as the cosmic center, Newton completed the job by showing that the same *physical* law held for both earthly and heavenly realms. Under this inspiration, geologists, assuming that the same laws also applied throughout time, showed Earth to be vastly older than the Biblical 6,000 years. This led directly to Darwin's theory of evolution, the most socially disturbing idea of modern science.

Though aspects of Newton's legacy will forever endure, the Newtonian mechanistic worldview, the bedrock of what we today call "classical physics," is challenged by modern physics. Nevertheless, this bedrock, our Newtonian heritage, still molds our commonsense view of the physical world and shapes our thinking in every intellectual sphere.

We now focus on five "commonsense" Newtonian stances, because we will soon show how quantum mechanics challenges each of them.

Determinism

Billiard balls are the physicist's much-loved model for determinism. If you know the position and velocity of a pair about to collide, with Newton's physics you can predict their position and velocity arbitrarily far into the future. Computers can calculate the future positions of a large number of colliding balls.

The same might be said, in principle, for the atoms bouncing around in a box of gas. Taking this idea all the way, to an "all-seeing eye" that knew the position and velocity of each atom in the universe at a given moment, the entire future of the universe would be apparent. The future of this Newtonian universe is, *in principle,* determined—whether or not anyone *knows* that future. The deterministic Newtonian universe is the Great Machine. The meshing gears of its clockworks move it inexorably on a predetermined course.

God then becomes the Master Clocksmith, the Great Engineer. Some went further: After making the completely deterministic machine, God had no role—he was a *retired* engineer. And moving from retirement to nonexistence was a small step.

Determinism gets personal: Are our seemingly free choices actually predetermined? According to Isaac Bashevis Singer, "You have to believe in free will. You have no choice." We have a paradox: Our free will conflicts with the determinism of Newtonian physics.

What about free will *before* Newton? No problem. In Aristotle's physics even a stone followed its individual inclination as it rolled down the hill in its own particular way. It is the determinism of Newtonian physics that presents the paradox.

It is, however, a benign paradox. Though we affect the physical world by our conscious free will, the only externally observable effect of conscious free will on the physical world comes about indirectly through our muscles that physically move things. Our consciousness can be seen as confined within our body.

Classical physics allows the tacit isolation of consciousness and its associated free will from the domain of physics' concern. There is mind, and there is matter. Physics deals with matter. With this divided universe, prequantum physicists could logically avoid the paradox. They could avoid it because the paradox arose only through the deterministic *theory,* not through any experimental demonstration. Thus, by limiting the scope of the theory, they could relegate free will and the rest of consciousness to psychology, philosophy, and theology. And that was their inclination.

We will see determinism challenged at the inception of quantum mechanics as Planck has electrons behaving randomly. A more profound challenge will be the intrusion of the conscious observer into the actual quantum experiment. No longer can the issue of free will be simply ruled out of physics by limiting the scope of the theory. It arises in the experimental demonstration. With quantum mechanics, the paradox of free will is no longer benign.

Physical Reality

Before Newton, explanations were mystical—and largely useless. If planets were pushed by angels, and rocks fell because of their innate desire for the cosmic center, if seeds sprouted craving to emulate their mature relatives, who could deny the influence of other occult forces? Or that the phases of the moon or incantations might be relevant? The flu, its full name "influenza," is so named because it was originally explained in terms of a supernatural *influence.*

By contrast, in the Newtonian worldview, Nature was a machine whose workings, though incompletely understood, need be no more mysterious than the clock whose gears are not seen. Acceptance of such a physically real world has become conventional wisdom. Though we may say the car "doesn't want to start," we expect the mechanic to find a physical explanation.

We raise the issue of "reality" because quantum mechanics challenges the classical view of it. But let's avoid a semantic misunderstanding. We're *not* talking of *subjective* reality, a reality that can differ from one person to the next. For example, we may say, "You create your own reality," meaning your *psychological* reality. We're talking of *objective* reality, realities we can all agree on, like that of a rock's position.

Philosophers have taken varied—even bizarre—stands on the nature of reality long before quantum mechanics. A conventional philosophical stance called "realism" has the existence of the physical world being independent of its observation. A more drastic version denies the existence of anything *beyond* physical objects. In this "materialist" view, consciousness, for example, should be completely understandable, in principle at least, in terms of the electrochemical properties of the brain. The tacit acceptance of such a materialist view, even its explicit defense, is not uncommon today.

Contrasting with the Newtonian realist or materialist attitude is the philosophical stance of "idealism" holding that the world that we perceive is not the actual world. Nevertheless, the actual world can be grasped with the mind.

An extreme idealist position is "solipsism." Here's its essence: *All* I ever experience are my own sensations. All I can know of my pencil is the sensation of yellow light on my retina and the pressure against my fingers. I cannot demonstrate that there is anything "real" about the pencil, or anything else, beyond my experienced sensations. (Appreciate that this paragraph is in first-person singular. The rest of you are, solipsistically speaking, just figments in my mental world.)

"If a tree falls in the forest, and no one hears it, is there any noise?" The realist answers: "Even if the air pressure variations we might experience as sound were heard by no one, they existed as a physically real phenomenon." The solipsist answers: "There wasn't even a tree unless I experienced it. Even then, only my conscious sensations *actually* existed." In this regard we quote philosopher Woody Allen: "What if everything is an illusion, and nothing exists? In that case, I've definitely overpaid for my carpet."

The intrusion of the conscious observer into the quantum experiment jolts our Newtonian worldview so dramatically that the philosophical issues of realism, materialism, idealism—even silly solipsism!—come up for discussion.

Separability

Renaissance science with its Aristotelian basis was replete with mysterious connectivities. Stones had an eagerness for the cosmic center. Acorns sought to emulate nearby oaks. Alchemists believed their personal purity influenced the chemical reactions in their flasks. By contrast, in the Newtonian worldview a hunk of matter, a planet or a person, interacts with the rest of the world *only* through the physically real forces impressed upon it by other objects. It is otherwise *separable* from the rest of the universe. Except for impressed physical forces, an object has no "connectedness" with the rest of the universe.

Physical forces can be subtle. For example, when a fellow, seeing a friend, adjusts his motion to meet her, the influencing force is carried by the light reflected from her and is exerted on rhodopsin molecules in his retina. On the other hand, we would have a *violation* of separability if a voodoo priest could stick a pin in a doll and thereby—without a connecting physical force—cause you pain.

Quantum mechanics includes instantaneous influences that violate separability. Einstein derided these as "voodoo forces." However, actual experiments now demonstrate that they do indeed exist.

Reduction

Often implicit in viewing the world as comprehensible is the reductionist hypothesis: that a complex system can—in principle, at least—be explained in terms of, or "reduced" to, its simpler parts. The working of an automobile engine, for example, can be explained in terms of the pressure of the burning gasoline pushing on the pistons.

Explaining a psychological phenomenon in terms of its biological basis would be the reduction of an aspect of psychology to biology. ("There is more of gravy than of grave to you," said Scrooge to Marley's ghost as he reduced his dream to a digestive problem.)

A chemist might explain a chemical reaction in terms of the physical properties of the involved atoms, something feasible today in simple cases. This would be reducing a chemical phenomenon to physics.

We can think of a hierarchy going from psychology to physics, which is firmly based on empirical facts. Scientific explanations are generally reductionist, moving toward more general basic principles. Though one moves in that *direction,* it is usually only by small steps. We will always need general principles specific to each level.

The classic example of a violation of reductionist ideas is the "vital force"

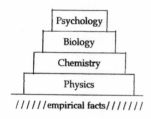

Figure 4.4 Hierarchy of scientific explanation

once proposed to account for life processes. Life supposedly emerged at the biological level without an origin in chemistry or physics. Such vitalist thinking led nowhere and, of course, has no standing in today's biology.

In studies of consciousness, reduction sparks controversy. Some argue that once the electrochemical neural correlates of consciousness are understood, there will be nothing left to explain. Others insist that the "inner light" of our conscious experience will elude the reductionist grasp, that consciousness is primary, and that new "psychophysical principles" will be needed. Quantum mechanics is claimed as evidence supporting this nonreductionist view.

A Sufficient Explanation

Newton was challenged to *explain* his force of gravity. A force transmitted through empty space, through *nothingness,* was a big pill to swallow.

Newton had a succinct response: *"Hypotheses non fingo"* ("I make no hypotheses"). He thus claims that a theory need do no more than provide consistently correct predictions. The *hypotheses non fingo* attitude arises again with quantum mechanics, where the challenge is to explain observations that force us to deny straightforward physical reality. It's an even bigger pill to swallow than a force transmitted through nothingness.

Beyond Physics by Analogy

In the decades following Newton, engineers learned to build the machines that launched the Industrial Revolution. Chemists moved beyond mystical alchemy, which for centuries had achieved almost nothing. Agriculture became scientific as understanding replaced folklore. Though the early workers in technology used almost no physics, the rapid advances they made required the Newtonian perspective: that discernable laws govern the physical world.

Newton's physics became the paradigm for all intellectual endeavors. Analogies with physics were big and bold. Auguste Comte invented the term "sociology" and referred to this science as "social physics," in which people were "social atoms" motivated by forces. The study of society had never previously been regarded as in any way scientific.

Pushing the analogy with Newtonian physics, Adam Smith argued for *laissez faire* capitalism by claiming that if people were allowed to pursue their own interests, an "invisible hand," a fundamental law of political economy, would regulate society for the general good.

Analogies are flexible. Karl Marx felt that he, not Adam Smith, had discovered the correct law. In *Das Kapital* he claimed to "lay bare the economic law of motion of modern society." With that law he predicted the communist future. By analogy with a mechanical system, he merely needed to know the initial condition, which, he thought, was the capitalism of his day. Thus, Marx's major work is a study of capitalism.

Analogies also arose in psychology. Sigmund Freud wrote: "It is the intention of this project to furnish us with a psychology which shall be a natural science. Its aim is to represent psychical processes as quantitatively determined states of specific material particles. . . ." Newtonian enough? As a later example, consider B. F. Skinner's declaration: "The hypothesis that man is not free is essential to the application of the scientific method to the study of human behavior." He explicitly denies free will, polemically adopting materialism and Newtonian determinism.

The appeal of such approaches in the social sciences has cooled. Workers in such complex areas are today more aware of the limitations of a method that works well for simpler physical situations. But the *broader* Newtonian perspective, the seeking of general principles that are then subject to empirical testing, is the accepted mode.

Being explicit about our Newtonian heritage helps us, personally, to appreciate the challenge that quantum mechanics poses to that worldview, one from which we can hardly escape.

5

All the Rest of Classical Physics

There is nothing new to be discovered in physics now. All that remains is more and more precise measurement.

—Lord Kelvin (in 1894)

Six years after Kelvin made this claim, he hedged: "Physics is essentially complete: There are just two dark clouds on the horizon." He picked the right clouds: One hid relativity; the other, quantum mechanics. But before we look behind them, we need to describe a bit more of the nineteenth-century physics we today call "classical." One doesn't need much technical detail to come to grips with the quantum enigma—we will describe interference, which demonstrates something to be a wave. We will need the concept of electric field—light is a rapidly varying electric field, and it is with light that the quantum enigma first arose. We will also talk of energy and its "conservation," its unchanging totality. And we will briefly tell of Einstein's theory of relativity—its well-confirmed but hard-to-believe predictions are good psychological practice for the "impossible-to-believe" implications of quantum theory.

The Story of Light

Newton decided that light was a stream of tiny particles. He had good arguments: Like objects obeying his universal equation of motion, light travels in straight lines unless it encounters something that might exert a force on it. His particles of light scatter widely when they hit something irregular but bounce off a mirror at the same angle they hit, as would a tennis ball. Newton even ascribes color to the sizes of these bodies:

> Are not the Rays of Light very small Bodies emitted from shining Substances? For such Bodies will pass through uniform Mediums in right Lines without bending into the Shadow, which is the Nature of the Rays of Light. . . . Nothing more is requisite for producing all the variety of Colours, and degrees of Refrangibility, than that the Rays of Light be Bodies of different sizes. . . .

Actually, Newton was conflicted. He investigated a property of light we now call "interference," a phenomenon uniquely characteristic of waves. Nevertheless, he came down strongly in favor of particles. Waves seemed to require a medium in which to propagate, and this medium would impede the motion of the planets that his universal equation of motion seemed to deny. As he put it:

> And against filling the Heavens with fluid Mediums, unless they be exceeding rare, a great Objection arises from the regular and very lasting Motions of the Planets and Comets in all manner of Courses through the Heavens. . . . [T]he Motions of the Planets and Comets being better explain'd without it. . . . [S]o there is no evidence for its Existence, and therefore it ought to be rejected. And if it be rejected, the Hypotheses that Light consists in Pression or Motion, propagated through such a Medium, are rejected with it.

Other scientists proposed wave theories of light, but the overwhelming authority of Newton meant that his "corpuscular theory," that light is a hail of little bodies, dominated for more than 100 years. The Newtonians were more sure of Newton's corpuscles than was Newton—until about 1800.

Thomas Young was a precocious child who, reportedly, read fluently at the age of two. He was educated in medicine, earned his living as a physician, and was an outstanding translator of hieroglyphics. But his main interest was physics. At the beginning of the nineteenth century, Young provided the convincing demonstration that light was a wave.

On glass made opaque with soot, Young scribed two closely spaced parallel lines. Light shining through these two slits onto a wall or reflecting screen produced a pattern of bright and dark bands we call an "interference pattern." Interference is the conclusive demonstration of wave behavior. Its explanation is central to the quantum enigma: that observation creates reality. This is the only classical physics we present in any detail—bear with us.

We can picture a "wave" as a moving series of peaks and valleys, or crests and troughs. Such crests and troughs can, for example, be seen through the flat side of an aquarium as ripples on the water surface. Another way to depict waves is the bird's-eye view, where we draw lines to indicate the crests. Waves on the ocean seen from an airplane look like this. We'll show waves both ways.

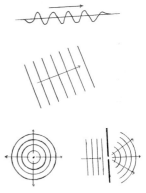

Waves from a small source, a pebble dropped into the water, for example, spread in all directions. Similarly, light from a tiny glowing object spreads in all directions. By the same token, light coming through a narrow slit to a screen illuminates the screen rather uniformly.

Light coming through two closely spaced slits might be expected to illuminate the screen twice as brightly. That would certainly be the expectation if light were a stream of little particles. But when Thomas Young shined

Figure 5.1 Views of waves

light through his two slits, he saw bands of brightness and darkness—a stream of particles could not account for this.

The explanation: At a central place on the screen (point A in figure 5.2), light waves from the top slit travel the same distance as light waves from the bottom slit. Therefore, crests from one slit arrive together with crests from the other. The crests add to produce more brightness than would exist if only one slit were open.

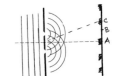

But to reach a place above the central location on the screen (point B in figure 5.2), waves from the bottom slit must travel farther than waves from the top slit. Therefore, at point B, crests from the bottom slit arrive later than crests from the top slit. If the extra

Figure 5.2 Interference in the double-slit experiment

distance of travel is just equal to the distance between a crest and a trough of the wave, crests from the bottom slit would always arrive at the same time as troughs from the top slit. Crests and troughs arriving together cancel each other to produce dark. With light being a wave, light plus light can add up to dark.

At a place yet farther up the screen (point C in figure 5.2), there will be another bright band, because at that place the difference in the distance from the two slits is just equal to the distance between crests (a wavelength). Thus, once again, crests from one slit arrive with crests from the other. Continuing up the screen, bright and dark bands will alternate as waves from the two slits alternately reinforce and cancel each other to form the interference pattern. "Interference" is actually a misnomer. Waves from the two slits are not interfering with each other; they just add and subtract, like deposits and withdrawals from a bank account.

If you think about the geometry a bit, you can see that the greater the spacing between the slits, the smaller is the spacing between the bright bands of the interference pattern. But the essential point to remember is that the interference pattern spacing depends on the slit spacing. Thus, an interference pattern shows that the light waves reaching each point on the screen must come from *both* slits.

Were light a stream of particles, there would be no interference pattern. Little bullets, each coming through one slit or the other, could not cancel each other to produce a pattern depending on the slit separation.

Is Young's argument airtight? Probably not. When Young presented it, it was hotly disputed. Young's English colleagues were strong in the Newtonian particle school of thought. Moreover, wave ideas were favored by *French* scientists and were rejected partly for that reason. But before long, further experiments over-whelmed objections to wave theory.

The Electromagnetic Force

A piece of silk rubbed on glass is attracted to the glass but repelled by another piece of silk rubbed on glass. Such "electric charge," seen when different materials were rubbed together, was long known. The crucial step in understanding it was the bright idea of Benjamin Franklin. He noticed that when any two electrically charged attracting bodies came into contact, the attraction lessened. He realized that attracting charged bodies canceled each other's charge.

Cancellation is a property of positive and negative numbers. Franklin there-fore assigned algebraic signs, positive (+) and negative (−), to charged objects. Bodies with charges of opposite sign attract each other. Bodies with charges of the same sign repel each other.

(Franklin's work on electricity is in good part responsible for the existence of the United States. As ambassador to France, it was not just Franklin's wit, charm, and political acumen, but his stature as a scientist, that allowed him to recruit the French aid that was so crucial to the success of the American Revolution.)

We now know that atoms have a positively charged nucleus made up of positively charged protons and uncharged neutrons. Electrons, each with a nega-tive charge equal in magnitude to that of a proton, surround the nucleus. The

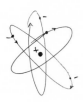

number of electrons in an atom is equal to the number of pro-tons, so the atom as a whole is uncharged. When two bodies are rubbed together, it is the electrons that move from one to the other.

A glass rod that is rubbed with a silk cloth, for example, becomes positively charged because electrons in the glass are less tightly bound than those in the silk. Therefore, some elec-trons move from the glass to the silk. The silk, now having more electrons than protons, is negatively charged and is at-tracted to the positively charged glass. Two negatively charged pieces of silk would repel each other.

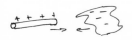

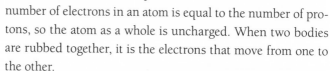

Figure 5.3 Positive and negative charges

Figure 5.4 Michael Faraday. Courtesy Stockton Press

A simple formula, Coulomb's law, tells us the strength of the electric force that one charged body (or "charge") exerts on another. With it you can calculate the forces in any arrangement of charges. That seemed to be the whole story of electric force—there was nothing more to say, or so thought most physicists in the early nineteenth century.

But Michael Faraday found the electric force puzzling. Let's back up a bit. At the age of fourteen, Michael Faraday, the son of a blacksmith, was apprenticed to a bookbinder. Faraday, a curious fellow, was fascinated by some popular science lectures by Sir Humphrey Davy. He took careful notes, bound them into a book, presented them to Sir Humphrey, and asked for a job in his laboratory. Though hired as a menial assistant, Faraday was soon allowed to try some experiments of his own.

How, Faraday wondered, could one body cause a force on another through empty space? That the mathematics of Coulomb's law correctly predicted what you would observe did not satisfy him. He therefore postulated that a charge creates an electric "field" in the space around itself, and it is this physical field that exerts forces on other charges. Faraday represented his field by lines emanating

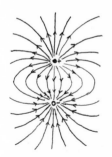

from a positive charge and going into a negative charge. Where the lines were most dense, the force the field exerted would be the greatest.

Most scientists, claiming that Coulomb's law said it all, considered Faraday's field concept to be superfluous. Faraday's ignorance of mathematics, they noted, required him to think in pictures; abstract thinking was no doubt difficult for this young man from the "lower classes." The field concept was ridiculed as "Faraday's mental crutch."

Figure 5.5 Electric field around two charges

Actually, Faraday went further and assumed that the field due to a charge takes time to propagate. If, for example, a positive and a nearby negative charge of equal magnitude were brought together to cancel each other, the field would disappear in their immediate neighborhood. But it seemed unlikely to Faraday that the field would disappear everywhere immediately.

The remote field would, he thought, exist for a while even when the charges that created it canceled each other and no longer existed. If true, the field would be a physically real thing in its own right.

Moreover, Faraday reasoned, if two equal and opposite charges were repeatedly brought together and separated, an alternating electric field would propagate from this oscillating pair. Even if they stopped oscillating and just canceled each other, the oscillating field would continue to propagate outward.

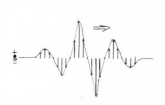

Faraday's intuition was sound. A few years later James Clerk Maxwell, picking up Faraday's field idea, devised a set of four equations that encompassed all electric and magnetic phenomena. We call them "Maxwell's equations." His striking prediction was the existence of waves of electric field propagating along with waves of magnetic field—"electromagnetic waves." Maxwell noticed that the speed of such waves was exactly what had been measured for light. He therefore proposed that light was an electromagnetic wave. This was in fact demonstrated soon after his death.

Figure 5.6 An oscillating electric field

As Faraday had predicted, the jiggling of charges produces electromagnetic radiation. The frequency of the jiggling is the frequency of the wave produced. Higher frequency jiggling produces violet and ultraviolet light; lower frequencies produce red and infrared.

Today the most fundamental theories in physics are formulated in terms of fields—Faraday's "mental crutch" is a pillar upon which all of physics now rests.

The electric force—short for electromagnetic force—is the only force we need talk of in this book. Along with gravity, it is the only force we normally experience. (Though all bodies exert gravitational forces on each other, it is only significant when at least one of the bodies is very massive, such as a planet.) The forces between atoms are essentially electrical.

When we touch someone, the pressure of our touch is an electric force. The electrons in the atoms of our hand repel the electrons in the atoms of the other person. Reach out and touch someone by telephone, and it is the electric force that carries the message over the wires and through space. The atoms making up solid matter are held together by electric forces. Electric forces are responsible for all of chemistry and therefore underlie all biology. We see, hear, smell, taste, and touch with electric forces. The processes in our brains are electrochemical, therefore ultimately electrical.

Is our thinking, our consciousness, ultimately to be explained wholly in terms of the electrochemistry taking place in our brain? Is our feeling of being conscious "merely" a manifestation of electrical forces? Some believe so. Others claim there is more to consciousness than electrochemistry—that's an issue for us to explore later.

There are forces in nature besides gravity and the electromagnetic force. But, it seems, only two others: the so-called "strong force" and "weak force." They both involve interactions of the particles making up the atomic nucleus (and objects created for an instant in high-energy particle collisions). They exert essentially no effect beyond the dimensions of the atomic nucleus. They'll not be important to us in this book.

Energy

Energy is a concept pervading physics, chemistry, biology, and geology, as well as technology and economics. Wars have been fought over the chemical energy stored in oil. The crucial aspect of energy is that, though its form may change, the total amount of energy stays constant. That fact, the "conservation of energy," is the first law of thermodynamics. But what *is* energy? We'll define it by pointing to energy in several of its different forms.

First of all, there is energy of motion. The larger the mass and the speed of a moving object, the larger its "kinetic energy." Energy due to the motion of objects is kinetic energy.

The farther a rock falls, the faster it goes and the larger its kinetic energy. A rock held at a certain height has the potential of gaining a certain speed. It has a gravitational "potential energy," which is larger for a larger mass or a larger height. The sum of a rock's kinetic and potential energy, its *total* energy, remains constant as the rock falls. This is an example of the conservation of energy.

Of course, after the rock hits the ground, it has zero kinetic energy and zero potential energy. As it contacts the ground, the energy of the rock itself was not conserved. But total energy is conserved. On impact, the rock's energy is given to internal random motion of the atoms of the ground and those of the rock. Those atoms now jiggle about with greater agitation. The haphazard motion of these atoms is the microscopic description of thermal energy (heat). Where the rock hit, the ground is warmer. The energy imparted to the jiggling atoms is just equal to the energy the rock lost on impact.

Although the total energy is conserved when the rock stops, the energy available for use decreases. The kinetic energy of falling rocks, or falling water, could, for example, be used to turn a wheel. But once energy goes over to the random motion of atoms, it is unavailable to us except as thermal energy. Moreover, the second law of thermodynamics tells us that in any action some energy becomes unavailable. When we're enjoined for environmental reasons to "conserve energy," we're being asked to conserve available energy.

There is only one kind of kinetic energy, but there are many kinds of potential energy. The energy of that rock held at some height is gravitational potential energy. A compressed spring or a stretched rubber band has elastic potential energy. The elastic energy of the spring can be converted to kinetic energy in projecting a rock upward.

When a positive and a negative electrical charge are held apart from each other, those charges have electrical potential energy. If released, they would fly toward each other with increasing speed and kinetic energy. In an atom, the electrons orbiting the nucleus have both kinetic and potential energy.

The chemical energy of a bottle of hydrogen and oxygen molecules is greater than the energy those molecules would have if they were bound together as water at the same temperature. Should a spark ignite that hydrogen–oxygen mixture, the greater energy would appear as kinetic energy of the resulting water molecules. The water vapor would therefore be hot. The chemical energy stored in the hydrogen–oxygen mixture would have become thermal energy.

Nuclear energy is analogous to chemical energy, except that the forces involved between the protons and the neutrons that make up the nucleus include nuclear (strong and weak) forces as well as electrical forces. A uranium nucleus has a greater total energy than do the fission products it breaks into. That greater energy becomes the kinetic energy of the fission products. That kinetic energy is thermal energy and can be used in a power reactor to make steam to turn turbines that turn generators to produce electric power. It can also become a bomb.

When light is emitted from a glowing hot body, energy goes to the electromagnetic radiation field, and the glowing body cools, unless it is supplied with additional energy. When a single atom emits light, it goes to a state of lower energy.

How many forms of energy are there? That depends on how you count. Chemical energy is, for example, ultimately electrical energy, though it is usually convenient to classify it separately. There may be forms of energy we don't yet know about. Just a few years ago it was discovered that the expansion of the universe is not slowing down, as was generally believed—it's accelerating. The vast amount of energy causing this acceleration has a name, "dark energy," but there is still more mystery about it than understanding.

What about "psychic energy"? Physics can claim no patent on the word "energy." It was used long before being introduced into physics in the early nineteenth century. If "psychic energy" could be converted into an energy treated by physics, it would be a form of the energy we're talking about. There is, of course, no generally accepted evidence for that.

Relativity

> Alice laughed. "There's no use trying," she said: "one *can't* believe impossible things."
>
> "I daresay you haven't had much practice," said the Queen. "When I was your age, I always did it for half-an-hour a day. Why sometimes I've believed as many as six impossible things before breakfast."
>
> —Lewis Carroll, *Through the Looking Glass*

When light became accepted as being a wave, it was assumed that something had to be waving. Electric and magnetic fields would be distortions in this waving medium. Since material bodies moved through it without resistance, it was ethereal and was called the "ether." It presumably pervaded the universe since we receive light from the stars. Motion with respect to this ether would define an *absolute* velocity, something not meaningful without ether as a stationary "hitching post" in the universe.

In the 1890s Albert Michelson and Edward Morley set out to determine how fast our planet was moving through the universal ether. A boat moving in the same direction as the waves sees the waves pass more slowly than when the boat moves in the direction opposite the waves. From the difference in these two wave speeds, one can determine how fast the boat is moving on the water. This is essentially the experiment Michelson and Morley did with light waves.

To their surprise, Earth seemed not to be moving at all. At least, they measured the speed of light to be the same in all directions. Ingenious attempts to untangle this result with electromagnetic theory failed. Albert Einstein took a different tack and cut the Gordian knot. He postulated the observed fact: that the speed of light is the same no matter how fast the observer moves. He took it as a new law of Nature. Two observers, though moving at different speeds, would each measure the same light beam to be passing them at the same speed. The speed of light in a vacuum is therefore a universal constant, called "c."

In that case, an absolute velocity could not be measured. Any observers, whatever their constant velocity, could consider themselves at rest. There is no absolute velocity; only *relative* velocities are meaningful—hence, the "theory of relativity."

With just simple algebra, Einstein deduced further testable predictions from his postulate. The prediction most important to us in this book is that no object, no signal, no information, can travel faster than the speed of light. Another prediction is that mass is a form of energy and can be converted into other forms of energy. It's summarized as $E = mc^2$. Both of these predictions have been confirmed, sometimes dramatically.

The prediction that is hardest to believe is that the passage of time is relative: Time passes more slowly for a fast-moving object than it does for something at rest.

Suppose a twenty-year-old woman travels to a distant star in her superfast rocket ship, leaving her twin brother on Earth for thirty years. On her return, her brother, having aged thirty years, is now a middle-aged fifty. She, for whom time passed more slowly at her speed of, say, ninety-five percent that of light, has aged only ten years. She would be a relatively young thirty. The traveler is twenty years younger than her stay-at-home twin in every physical and biological sense.

This "twin paradox" was raised early on as a supposed refutation of Einstein's theory. Could she not have considered herself at rest and her brother to have taken the speedy trip? He would then be younger than she. The theory, it was claimed, was inconsistent. Not so—the situation is not symmetric. Only observers moving at constant velocity (constant speed in a constant direction) can consider themselves at rest. That could not be true for the traveler, who had to turn around, to accelerate, at the distant star in order to return home.

While it is not technically feasible to build rocket ships to move people at near light speeds, relativity theory has been extensively tested and confirmed. Most tests have been with subatomic particles. But the theory has also been checked by comparing accurate clocks flown around the world with clocks that stayed home. On their return, the traveling clocks were "younger." They recorded a bit less time—by precisely the predicted amount. The validity of the theory is so well established today that only an extremely challenging test would be warranted. If you read about a test of "relativity," it is likely the test is of the theory of *general* relativity, Einstein's theory of gravity. The full name of the theory we're talking here about is the theory of *special* relativity.

It is hard to believe the strange things that Einstein's relativity theory tells us. That one could, in principle, become older than one's mother, for example. But accepting the established fact that moving systems age less is good practice for believing the far stranger things that quantum mechanics tells.

We're now ready to start talking about those strange things.

Hello Quantum Mechanics

The universe begins to look more like a great thought than a great machine.

—Sir James Jeans

As the nineteenth century ended, the search for Nature's basic laws seemed close to its goal. There was a sense of a task accomplished. Physics presented an orderly scene that fit the proper Victorian mood of the day.

Objects both on Earth and in the heavens behaved in accord with Newton's laws. So, presumably, did atoms. The nature of atoms was unclear. But to most scientists the rest of the job of describing the universe seemed a filling in of the details of the Great Machine.

Did the determinism of Newtonian physics deny "free will"? Physics would leave such fuzzy questions to philosophy. Defining the territory that physicists considered their own seemed reasonable and straightforward. There was little to motivate a search for deeper meaning behind Nature's laws. But this intuitively sensible worldview could not account for what physicists soon saw in their laboratories, what at first seemed like "details."

Classical physics explains the world quite well; it's just the details it can't handle. Quantum physics handles the details perfectly; it's just the world it can't explain.

Quantum physics does not *replace* classical physics the way the sun-centered solar system replaced the earlier view with Earth as the cosmic center. Rather, quantum physics *encompasses* classical physics as a special case. Classical physics is usually an extremely good approximation for behavior of objects that are much larger than atoms. But if you dig deeply enough into any natural phenomenon—physical, chemical, biological, or cosmological—you hit quantum mechanics.

Quantum theory has been subject to challenging tests for eight decades. No prediction by the theory has ever been shown wrong. It is the most battle-tested theory in all of science—it has no competitors. Nevertheless, if you take the implications of the theory seriously, you confront an enigma. The theory seems to tell us that the reality of the physical world depends on our observation of it. This is surely almost impossible to believe.

Being hard to believe presents a problem: Told something hard to believe, a likely response is: "I don't understand." There is also a tendency to reinterpret what is said to make it seem reasonable. Don't use believability or reasonableness as a test of comprehension. But here's one test: Niels Bohr, a founder of quantum theory, claimed that unless you're shocked by quantum mechanics, you have not understood it.

Though our presentation may be novel, the experimental facts we describe and quantum theory explanations we offer are standard and undisputed. We step beyond that firm ground when we explore the *interpretation* of the theory and thus physics' encounter with consciousness. The deeper meaning of quantum mechanics is in dispute.

It does not require a technical background to move to the frontier where physics joins issues that seem beyond physics and where physicists cannot claim unique competence. Once there, you can take sides in the debate.

6

How the Quantum Was Forced on Physics

It was an act of desperation.

—Max Planck

Physics courses are rarely presented historically. The introductory course in quantum mechanics is the exception. For students to see why we accept a theory so violently in conflict with common sense, they must see how physicists were dragged from their nineteenth-century complacency by the brute facts observed in their laboratories.

The Reluctant Revolutionary

In the final week of the nineteenth century, Max Planck suggested something outrageous—that the most fundamental laws of physics were violated. This was the first hint of the quantum revolution, that the worldview we now call "classical" had to be abandoned.

Max Planck, son of a distinguished professor of law, was careful, proper, and reserved. His clothes were dark and his shirts stiffly starched. Raised in the strict Prussian tradition, Planck respected authority, both in society and in science. Not only should people rigorously obey the laws, so should physical matter. Not your typical revolutionary.

In 1875, when young Max Planck announced his interest in physics, the chairman of his physics department suggested he study something more exciting. Physics, he said, was just about complete: "All the important discoveries have already been made." Undeterred, Planck completed his studies in physics and plugged away for years as a *Privatdozent,* an apprentice professor, receiving only the small fees paid by students attending his lectures.

Planck chose to work in the most properly lawful area of physics, thermo-

Figure 6.1 Max Planck

dynamics, the study of heat and its interaction with other forms of energy. His solid but unspectacular work eventually won him a professorship. His father's influence is said to have helped.

A nagging unexplained phenomenon in thermodynamics was thermal radiation: the spectrum, the colors, of the light given off by hot bodies. (The problem was one of Kelvin's two "clouds.") Planck set about to solve it.

What's the problem? That a hot poker should glow seems obvious. Although at the turn of the century the nature of atoms, even the existence of atoms, was unclear, electrons had just been discovered. Presumably these little charged particles jiggled in a hot body and therefore emitted electromagnetic radiation. This light seemed important to understand as a fundamental aspect of Nature because it was the same no matter what material it came from.

The radiation one observed seemed reasonable. As a piece of iron gets hotter, its electrons should shake harder and, presumably, at a higher rate, meaning at a higher frequency. Therefore, the hotter the metal, the brighter and higher fre-

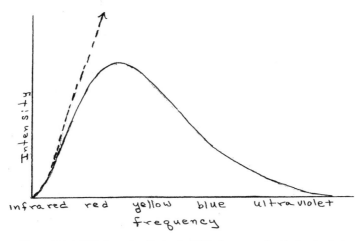

Figure 6.2 6,000° thermal radiation (solid line) compared with the classical prediction (dashed line)

quency the glow. As it gets hotter, its color goes from the invisible infrared, to a visible red, to orange, and eventually the metal becomes white hot as the emitted light covers the entire visible frequency range.

Since our eyes can't see frequencies above the violet, superhot objects, which emit mostly in the ultraviolet, appear bluish. Materials on Earth vaporize before they get hot enough to glow blue, but we can look up at hot blue stars. Even cool objects "glow," though weakly and at low frequencies. Bring your palm close to your cheek and feel the warmth from the infrared light your hand emits. The sky shines down on us with invisible microwave radiation left over from the flash of the Big Bang.

In figure 6.2, we sketch the actual intensity of radiation from the sun's 6,000° C surface at different frequencies, which we just label as colors. An object hotter than the sun emits more light at all frequencies, and its maximum intensity is at a higher frequency. But the intensity always drops at very high frequencies.

The dashed line is the problem—it is the intensity calculated with the laws of physics accepted in 1900. It worked well in the infrared. But at higher frequencies, classical physics not only gave a wrong answer, it gave a ridiculous answer: It predicted a forever increasing light intensity at frequencies beyond the ultraviolet.

Were this true, every object would instantaneously lose its heat by radiating a burst of energy at frequencies beyond the ultraviolet. This embarrassing deduction was derided as the "ultraviolet catastrophe." But no one could say where the seemingly sound reasoning went wrong.

Max Planck struggled for years to derive a formula that fit the experimental data. In frustration, he decided to work the problem backward. He would first try to guess a formula that agreed with the data and then, with that as a hint, try to develop the proper theory. In a single evening, studying the data others had given him, he found a fairly simple formula that worked perfectly.

If Planck put in the temperature of the body, his formula gave the correct radiation intensity at every frequency. His formula needed a "fudge factor" to make it fit the data, a number he called "h." We now call it "Planck's constant" and recognize it as a fundamental constant of Nature, like the speed of light.

Figure 6.3 Energy loss by charged particle according to classical physics

With his formula as a hint, Planck sought to explain thermal radiation in terms of the fundamental principles of physics. In the most straightforward models, an electron, though bound to its parent atom, would start vibrating if it were bumped by a jiggling neighboring atom in a hot metal. This little charged particle would then gradually lose its energy by emitting light. We plot such an energy loss in figure 6.3. In a similar fashion, a pendulum bob on a string, or a child on a swing, given a shove, would continuously lose energy to air resistance and friction.

However, every description of the electron radiating energy according to the physics of the day led to the same crazy prediction, the ultraviolet catastrophe. After a long struggle, Planck ventured an assumption that absolutely violated the universally accepted principles of physics. At first, he didn't take it seriously. He later called it "an act of desperation."

Max Planck assumed an electron could radiate energy only in chunks, in "quanta" (the plural of quantum). Moreover, each quantum would have an energy equal to the number h in his formula times the frequency of vibration of the electron.

Behaving this way, an electron would vibrate for a while at constant energy. That is, this electric charge would vibrate without losing energy to radiation. Then, randomly, and *without cause,* without an impressed force, it would suddenly lose a quantum of energy, radiating it as a pulse of light. (Electrons would

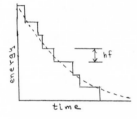

Figure 6.4 Energy loss by charged particle according to Planck

also gain their energy from the hot atoms by such "quantum jumps.") In figure 6.4 we plot an example of such energy loss in sudden steps. The dashed line repeats the classically predicted, gradual energy loss.

Planck was allowing the electrons to violate both the laws of electromagnetism and Newton's universal equation of motion. Only by this wild assumption could he get the formula he had guessed, the formula that correctly described thermal radiation.

If this quantum-jumping behavior is indeed a law of Nature, it should apply to everything. Why, then, do we see the things around us behaving smoothly? Why don't we see children on swings suddenly change their swinging motion in quantum jumps? It's a question of numbers, and *h* is an extremely small number.

Not only is *h* small, but since the frequency of a child moving back and forth on a swing is much lower than the frequency at which an electron vibrates, the quantum steps of energy (*h* times frequency) are vastly smaller for the child. And, of course, the total energy of a swinging child is vastly larger than that of an electron. Therefore, the number of quanta involved in the child's motion is vastly, vastly greater than the number involved in the motion of the electron. A quantum jump, the change in energy by a single quantum, is thus far too small to be seen for the child on a swing.

But back to Planck's day and the reaction to the solution he proposed for the thermal radiation problem: His formula fit the experimental data well. But his explanation seemed more confounding than the problem it presumed to solve. Planck's theory seemed silly. No one laughed, at least not in public—Herr Professor Planck was too important a man for that. His quantum-jumping suggestion was simply ignored.

Physicists were not about to challenge the fundamental laws of mechanics and electromagnetism. Even if the classical laws gave a ridiculous prediction for the light emitted by glowing bodies, these basic principles seemed to work everyplace else. And they made sense. Planck's colleagues felt a reasonable solution would eventually be found. Planck himself agreed and promised to seek one. The quantum revolution arrived with an apology, and almost unnoticed.

In later years, Planck even came to fear the negative *social* consequences of quantum mechanics. Freeing the fundamental constituents of matter from the rules of proper behavior might seem to free people from responsibility and duty. The reluctant revolutionary would have liked to cancel the revolution he sparked.

The Technical Expert, Third Class

His parents worried about mental retardation when young Albert Einstein was slow in starting to talk. Later, though he became an avid and independent student of things that interested him, his distaste for the rote instruction of the *Gymnasium* (high school) led to his not doing well. Asked to suggest a profession that Albert might follow, the headmaster confidently predicted: "It doesn't matter; he'll never make a success of anything."

Einstein's parents left Germany for Italy after the family electrochemical

Figure 6.5 Albert Einstein. Courtesy California Institute of
Technology and the Hebrew University of Jerusalem

business failed. The new business in Italy fared little better. Young Einstein was
soon on his own. He took the entrance exam to the Zurich Polytechnic Institute
but did not pass. He was finally admitted the next year. On graduation, he was
unsuccessful in trying for a position as *Privatdozent*. He had the same luck in applying for a teaching job at the *Gymnasium*. For a while Einstein supported himself as a tutor for students having trouble with high school. Eventually, through
a friend's influence, he got a job in the Swiss patent office.

His duties as Technical Expert, Third Class, were to write summaries of patent applications for his superiors to use in deciding whether an idea warranted a
patent. Einstein enjoyed the work, which did not take his full time. Keeping an
eye on the door in case a supervisor came in, he worked on his own projects.

Initially, Einstein continued on the subject of his doctoral thesis, the statistics
of atoms bouncing around in a liquid. This work soon became the best evidence
for the atomic nature of matter, something still debated at the time. Einstein was
struck by a mathematical similarity between the equation for the motion of atoms

and Planck's radiation law. He wondered: Might light be not only mathematically like atoms, but also physically like atoms?

If so, might light, like matter, come in compact lumps? Perhaps the pulses of light energy emitted in one of Planck's quantum jumps did not expand in all directions as Planck assumed. Could the energy instead be confined to a small region? Might there be atoms of light as well as atoms of matter?

Einstein speculated that light is a stream of compact lumps, "photons" (a term that came later). Each photon would have an energy equal to Planck's quantum (Planck's constant times its frequency). Photons would be created when electrons emit light. Photons would disappear when light is absorbed.

Seeking evidence that his speculation might be right, Einstein looked for something that might display a granular aspect to light. It was not hard to find. The "photoelectric effect" had been known for almost twenty years. Light shining on a metal could cause electrons to pop out.

The situation was messy. Unlike thermal radiation, where a universal rule held for all materials, the photoelectric effect for each substance was different. Moreover, the data were inaccurate and not particularly reproducible.

But never mind the bad data. Spread out light *waves* shouldn't kick electrons out of a metal at all. Electrons are too tightly bound. While electrons are free to move about within a metal, they can't readily escape it. We can "boil" electrons out of a metal, but it takes a very high temperature. We can pull electrons out of a metal, but it takes a very large electric field. Nevertheless, dim light, corresponding to an extremely weak electric field, still ejects electrons. The dimmer the light, the fewer the electrons. But no matter how dim the light, some electrons were always ejected.

Einstein gleaned more information from the bad data. Electrons popped out with high energy when the light was ultraviolet or blue. With lower frequency yellow light, their energy was less. Red light usually ejected no electrons. The higher the frequency of the light, the greater the energy of the emitted electrons.

The photoelectric effect was just what Einstein needed. Planck's radiation law implied that light came in packets, quanta, whose energy was larger for higher frequency light. If the quanta were actually compact lumps, all the energy of each photon could be concentrated on a single electron. A single electron absorbing a whole photon would gain a whole quantum of energy.

Light, especially high-frequency light with its high-energy photons, could then give electrons enough energy to jump out of the metal. The higher the energy of the photon, the higher the energy of the ejected electron. For light below a certain frequency, its photons would have insufficient energy to remove an electron from the metal, and no electrons would be ejected.

Einstein said it clearly in 1905:

According to the presently proposed assumption the energy in a beam of light emanating from a point source is not distributed continuously over larger and larger volumes of space but consists of a finite number of energy quanta, localized at points of space which move without subdividing and which are absorbed and emitted only as units.

Assuming that light comes as a stream of photons and that a single electron absorbs all the energy of a photon, Einstein used the conservation of energy to derive a simple formula relating the frequency of the light to the energy of the ejected electrons. We plot it in figure 6.6. Photons with energy less than that binding the electrons into the material could not kick any electrons out at all.

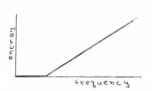

Figure 6.6 Energy of ejected electrons versus light frequency

A striking aspect of Einstein's photon hypothesis is that the slope of the straight line on this graph is just Planck's constant, h. Until this time, Planck's constant was just a number needed to fit Planck's formula to the observed thermal radiation. It appeared nowhere else in physics. Before Einstein's photon hypothesis, there was no reason to think the ejection of electrons by light had anything at all to do with the radiation emitted by hot bodies. This slope was the first indication that the quantum was universal.

Ten years after Einstein's work on the photoelectric effect, the American physicist Robert Millikan found that Einstein's formula in every case predicted "exactly the observed results." Nevertheless, Millikan called Einstein's photon hypothesis leading to that formula "wholly untenable" and called Einstein's suggestion that light came as compact particles "reckless."

Millikan was not alone. The physics community received the photon postulate "with disbelief and skepticism bordering on derision." Nevertheless, eight years after proposing the photon, Einstein had gained a considerable reputation as a theoretical physicist for many other achievements and was nominated for membership in the Prussian Academy of Science. Planck, in his letter supporting that nomination, felt he had to defend Einstein: "[T]hat he may sometimes have missed the target in his speculations, as, for example, in his hypothesis of light quanta, cannot really be held too much against him. . . ."

Even when Einstein was awarded the Nobel Prize in 1922 for the photoelectric effect, the citation avoided explicit mention of the then seventeen-year-old, but still unaccepted, photon. An Einstein biographer writes: "From 1905 to 1923, [Einstein] was a man apart in being the only one, or almost the only one, to take the light-quantum seriously." (We tell what happened in 1923 later in this chapter.)

Though the reaction of the physics community to Einstein's photons was, in a word, rejection, they were not just pig-headed. Light was proven to be a spread-out wave. Light displayed interference. A stream of discrete particles could not do that.

Recall our discussion of interference in chapter 5: Light coming through a single narrow slit illuminates a screen more or less uniformly. Open a second slit, and a pattern of dark bands appears whose spacing depends on the spacing of the two slits. At those dark places, wave crests from one slit arrive together with wave troughs from the other. Waves from one slit thus cancel waves from the other. Interference demonstrates that light is a wave.

Nevertheless, Einstein held that the photoelectric effect showed light to be a stream of photons—tiny compact bullets. But how could these tiny bullets produce the interference patterns seen with light?

In our previous chapter we mentioned that the argument that tiny bullets could not cause interference was not airtight. Might they not somehow deflect each other to form the bright and dark bands? That loophole in the argument has been closed. Interference can be seen with light so dim that only one photon is present at a time.

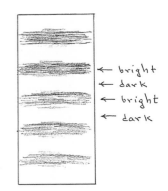

Figure 6.7
An interference pattern

Choosing to demonstrate interference, something explicable only in terms of waves, you could prove light to be a widely spread-out wave. However, by choosing a photoelectric demonstration, where a single electron absorbed a whole light quantum, you could prove light to be a stream of tiny compact objects. There seems to be an inconsistency. (Recall that something like this was seen in Neg Ahne Poc: Our visitor could choose to prove the couple was an entity spread over both huts, or he could choose to prove the couple was a concentrated entity in a single hut.)

Though the paradoxical nature of light disturbed Einstein, he clung to his photon hypothesis. He declared that a mystery existed in Nature and that we must confront it. He did not pretend to resolve the problem. And we do not pretend to resolve it here in this book. The mystery is still with us a hundred years later. The implications of our being able to choose to prove either of two contradictory things extend beyond physics. It's the quantum enigma. We will see far-out speculations being seriously proposed.

In 1906, the year after Albert Einstein discovered the quantum nature of light, firmly established the atomic nature of matter, and formulated the theory of relativity, he was promoted by the Swiss patent office to Technical Expert, *Second* Class.

Figure 6.8 Niels Bohr. Courtesy the American Institute of Physics

The Postdoc

Niels Bohr grew up in a comfortable and respected family that nurtured independent thought. His father, a professor of psychology at Copenhagen University, was interested in philosophy and science and encouraged those interests in his two sons. Niels's brother, Harald, eventually became an outstanding mathematician. Niels Bohr's early years were supportive. Unlike Einstein, he was never the rebel.

In college in Denmark, Bohr won a medal for some clever experiments with fluids. But we skip ahead to 1912 when, with his new Ph.D., Bohr went to England as a "postdoc," a postdoctoral student.

By this time the atomic nature of matter was generally accepted, but the atom's internal structure was unknown—actually, it was in dispute. Electrons, negatively charged particles thousands of time lighter than any atom, had been discovered a decade earlier by J. J. Thompson. An atom, being electrically neutral, must somewhere have a positive charge equal to that of its negative electrons, and that positive charge presumably had most of the mass of the atom. How were the atom's electrons and its positive charge distributed?

Thompson had made the simplest assumption: The massive positive charge uniformly filled the atomic volume and the electrons—one in hydrogen and almost 100 in the heaviest known atoms—were distributed throughout the positive background like raisins in a rice pudding. Theorists tried to calculate how various distributions of electrons might give each element its characteristic properties.

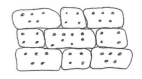

Figure 6.9 Thompson's rice pudding model of atoms

There was a competing model for the atom. Ernest Rutherford at the University of Manchester in England explored the atom by shooting alpha particles (helium atoms stripped of their electrons) through a gold foil. He saw something inconsistent with Thompson's uniformly distributed positive mass. About one alpha in 10,000 would bounce off at a large angle, sometimes even backward. The experiment was likened to shooting prunes through rice pudding—collisions with raisins could not knock a fast prune much off track. Rutherford concluded that his alpha particles were colliding with an atom's positive charge, and that almost all the atom's mass was concentrated in a small lump, a "nucleus."

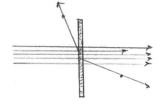

Figure 6.10 Rutherford's experiment with alpha particles

Why, however, did the negative electrons, attracted by the positive nucleus, not just fall into it? For the same reason that planets don't crash down into the sun: They orbit the sun. Rutherford decided that electrons orbited a small, massive, positive nucleus.

There was a problem with Rutherford's planetary model: instability. Since an electron is charged, it should radiate as it races around its orbit. Calculations showed

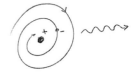

Figure 6.11 Instability of Rutherford's atomic model

that an electron should give off its energy as light and spiral down to crash into the nucleus in less than a millionth of a second.

Most of the physics community considered the instability in the planetary model a more serious problem than the rice pudding model's inability to explain the rare large-angle deflections of Rutherford's alpha particles. But Rutherford, a supremely confident fellow, knew his planetary model was basically right.

When the young postdoc Bohr arrived in Manchester, Rutherford assigned him the job of explaining how the planetary atom might be stable. Bohr's tenure in Manchester lasted only six months, supposedly because his support money ran out. But an eagerness to get back to Denmark to marry the beautiful Margrethe likely shortened his stay. While teaching at the University of Copenhagen in 1913, Bohr continued to work on the stability problem.

How he got his successful idea is not clear. But while other physicists were trying to understand how the quantum of energy and Planck's constant, h, arose from the classical laws of physics, Bohr took an "h okay!" attitude. He just accepted quantization as fundamental. After all, it worked for Planck, and it worked for Einstein.

Bohr wrote a very simple formula that said that "angular momentum," the rotational motion of an object, could exist only in quantum units. If so, only certain electron orbits were allowed. And, most important, he wrote his formula so that there was a smallest possible orbit. By fiat, Bohr's formula "forbid" an electron to crash into the nucleus. If his ad hoc formula was correct, the planetary atom was stable.

Without more evidence, Bohr's quantum idea would be rejected out of hand. But from his formula Bohr could readily calculate all the energies allowed for a single electron orbiting a nucleus, that is, for the hydrogen atom. From those energies he could then calculate the particular frequencies of light that could be emitted from hydrogen atoms electrically excited in a "discharge," something like a neon sign only with hydrogen instead of neon.

Those frequencies had been carefully studied for years, though Bohr was initially unaware of that work. Why only certain frequencies were emitted was a complete mystery. The spectrum of frequencies, unique to each element, presented a pretty set of colors. But were they any more significant than the particular patterns of a butterfly's wings? Now, however, Bohr's quantum rule predicted the frequencies for hydrogen with stunning accuracy—precise to parts in 10,000. But at this time, while Bohr had light *quanta* emitted by atoms, he, along with essentially all other physicists, still rejected Einstein's compact photon.

Some physicists nevertheless dismissed Bohr's theory as "number juggling." Einstein, however, called it "one of the greatest discoveries." And others soon came to agree. Bohr's basic idea was rapidly applied widely in physics and chemistry. No one understood why it worked. But work it did. And for Bohr that was the important thing. Bohr's pragmatic "h okay!" attitude toward the quantum brought him quick success.

Contrast Bohr's early triumph with his quantum ideas with Einstein's long remaining "a man apart" in his belief in the almost universally rejected photon. Watch how the early experiences of these two men is reflected in their lifelong friendly debate about quantum mechanics.

Figure 6.12 Louis de Broglie.
Courtesy the American Institute of Physics

The Prince

Louis de Broglie was *Prince* Louis de Broglie. His aristocratic family intended a career in the French diplomatic service for him, and young Prince Louis studied history at the Sorbonne. But after receiving an arts degree, he moved to theoretical physics. Before he could do much physics, World War I broke out, and de Broglie served in the French army at a telegraph station in the Eiffel Tower.

With the war over, de Broglie started work on his physics Ph.D., attracted, he says, "by the strange concept of the quantum." Three years into his studies, he read the recent work of the American physicist Arthur Compton. An idea clicked in his head. It led to a short doctoral thesis and eventually to a Nobel Prize.

Compton had, in 1923, almost two decades after Einstein proposed the photon, discovered, to his surprise, that when light bounced off electrons its frequency changed. This is not wave behavior: When a wave reflects from an object, each incident crest produces one other wave crest. The frequency of the wave therefore does not change in reflection from a stationary object. On the other hand, if Compton assumed that light was a stream of particles, *each with the energy of an Einstein photon,* he got a perfect fit to his data.

The "Compton effect" did it. Physicists now accepted photons. Sure, in certain experiments light displayed its spread-out wave properties and in others its compact particle properties. As long as one knew under what conditions each property would be seen, the photon idea seemed less troublesome than finding another explanation for the Compton effect. Einstein, however, still "a man apart," insisted a mystery remained, once saying: "Every Tom, Dick, and Harry thinks they know what the photon is, but they're wrong."

Graduate student de Broglie shared Einstein's feeling that there was a deep meaning to light's duality, being either extended wave or compact particles. He wondered whether there might be symmetry in Nature. If light was either wave or particle, perhaps matter was also either particle or wave. He wrote a simple expression for the wavelength of a particle of matter. This formula for the "de Broglie wavelength" of a particle is something every beginning quantum mechanics student quickly learns.

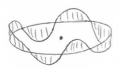

Figure 6.13 De Broglie's symmetry idea

The first test of that formula came from a puzzle that stimulated de Broglie's wave idea: If an electron in a hydrogen atom were a compact particle, how could it possibly "know" the size of an orbit in order to follow only those orbits allowed by Bohr's by-now-famous formula?

The lengths of violin string required to produce a given pitch are determined by the whole number of half-wavelengths of vibration that fit along the length of the string. Similarly, if the electron was a wave, the allowed orbits might be determined by a whole number of electron wavelengths that fit around the orbit's circumference. Applying this idea, de Broglie was able to derive Bohr's ad hoc quantum rule. (In the violin, it's the material of the string that vibrates. What vibrates in the case of the electron "wave" was then a mystery. It's become an even deeper one.)

Figure 6.14
Wavelengths around an electron orbit

It's not clear how seriously de Broglie took his conjecture. He certainly did not recognize it as advancing a revolutionary view of the world. In his own later words:

[H]e who puts forward the fundamental ideas of a new doctrine often fails to realize at the outset all the consequences; guided by his personal intuitions, constrained by the internal force of mathematical analogies, he is carried away, almost in spite of himself, into a path of whose final destination he himself is ignorant.

De Broglie took his speculation to his thesis adviser, Paul Langevin, famous for his work on magnetism. Langevin was not impressed. He noted that in deriving Bohr's formula de Broglie merely replaced one ad hoc assumption with another. And de Broglie's assumption, that electrons could be waves, seemed ridiculous.

Were de Broglie an ordinary graduate student, Langevin might have summarily dismissed his idea. But he was *Prince* Louis de Broglie. Aristocracy was meaningful, even in the French republic. So no doubt to cover himself, Langevin asked for a comment on de Broglie's idea from the world's most eminent physicist. Einstein replied that this young man has "lifted a corner of the veil that shrouds the Old One."

Meanwhile, there was a minor accident in the laboratories of the telephone company in New York. Clinton Davisson was experimenting with the scattering of electrons from metal surfaces. While Davisson's interests were largely scientific, the phone company was developing vacuum tube amplifiers for telephone transmissions, and for that the behavior of electrons striking metal was important.

Electrons usually bounced off a rough metal surface in all directions. But after the accident, in which a leak allowed air into his vacuum system and oxidized a nickel surface, Davisson heated the metal to drive off the oxygen. The nickel crystallized, essentially forming an array of slits. Electrons now bounced off in only a few well-defined directions. It was an interference pattern demonstrating the electron's wave nature. The discovery confirmed de Broglie's speculation that material objects could also be waves.

We opened this chapter with the first hint of the quantum in 1900. It was a hint largely ignored. We close it with physicists in 1923 finally forced to accept a wave–particle duality: A photon, an electron, an atom, a molecule—in principle any object—can be either compact or widely spread-out. You can show something to be either bigger than a breadbox or smaller than an atom. You can choose which of these two contradictory features to demonstrate. The physical reality of an object depends on how you *choose* to look at it.

Physics had encountered consciousness but did not yet realize it. Awareness of that contact came a few years later, after Schrödinger's discovery of the new universal law of motion. That discovery is the subject of our next chapter.

7

Schrödinger's Equation
The *New* Universal Law of Motion

*If we are still going to put up with these damn
quantum jumps, I am sorry that I ever had anything to
do with quantum theory.*

—Erwin Schrödinger

By the early 1920s physicists had accepted the fact that, depending on the experimental setup, matter as well as light could be displayed either as compact lumps or as widely spread-out waves. Few pretended to understand this seeming contradiction. The significance of this came a few years later with the Schrödinger equation. But Erwin Schrödinger wasn't looking for significance. He saw de Broglie's matter waves as a way to get rid of Bohr's "damn quantum jumps."

Erwin Schrödinger, the only child of a prosperous Viennese family, was an outstanding student. As an adolescent he became intensely interested in the theater and in art. Both were areas of rebellion against the bourgeois society of late nineteenth-century Vienna. Schrödinger himself rejected the Victorian morality of his upbringing. Throughout his life he channeled much energy into intense romances, his lifelong marriage notwithstanding.

After serving in the First World War as a lieutenant in the Austrian army on the Italian front, Schrödinger started teaching at the University of Vienna. About this time he embraced the Indian mystical teaching Vedanta but always kept this philosophical leaning apart from his physics. In 1927, just after his spectacular work in quantum mechanics, he was invited to Berlin University as Planck's successor. With Hitler's coming to power in 1933, Schrödinger, though not Jewish, left Germany. After visits to England and the United States, he incautiously returned to his native Austria to accept a position at the University of Graz. With Hitler's annexation of Austria, he was in trouble. His leaving Germany established his opposition to the Nazis. He escaped to Italy and spent the rest of his career at the School for Theoretical Physics in Dublin, Ireland.

Figure 7.1 Erwin Schrödinger. Courtesy the American Institute of Physics

A Wave Equation

Despite the successes of the early quantum theory, often based on Bohr's quantum rule, Schrödinger rejected a physics where electrons moved only in "allowed orbits" and then, without cause, abruptly jumped from one orbit to another. He was outspoken:

> You surely must understand, Bohr, that the whole idea of quantum jumps necessarily leads to nonsense. It is claimed that the electron in a stationary state of an atom first revolves periodically in some sort of an orbit without radiating. There is no explanation of why it should not radiate; according to Maxwell's theory, it must radiate. Then the electron jumps from this orbit to another one and thereby radiates. Does the transition occur gradually or suddenly? . . . And what laws determine its motion in a jump? Well, the whole idea of quantum jumps must simply be nonsense.

Schrödinger credits Einstein's "brief but infinitely far-seeing remarks" for calling his attention to de Broglie's speculation that material objects could display a wave nature. The idea appealed to Schrödinger. Waves might evolve smoothly from one state to another. Electrons would not need to orbit without radiating. He might get rid of Bohr's "damn quantum jumps."

Willing to amend Newton's laws to account for the quantum behavior of small objects, Schrödinger nevertheless wanted a description of the world that had electrons and atoms behaving reasonably. He would seek an equation governing waves of matter. It would be new physics, a guess that would have to be tested. Schrödinger would seek the *new* universal equation of motion. The old classical physics would be merely the good approximation for large objects.

From the position and motion of a tossed stone at one moment, Newton's law predicts its future position and motion. Similarly, from a wave's initial shape, a wave equation predicts its shape at any later time. It describes how the ripples spread from the spot where a tossed pebble hits the water, or how waves propagate on a taut rope.

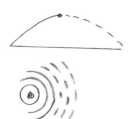

However, the single-wave equation that works for waves of water, light, and sound doesn't work for matter waves. Water, light, and sound waves move at the single speed determined by the medium in which the wave propagates. Sound, for example, moves at 330 meters per second in air. The wave

Figure 7.2 The path of a stone and the spreading of water ripples

equation Schrödinger sought had to allow matter waves to move at *any* speed because electrons, atoms—and baseballs—move at any speed.

The breakthrough came during a mountain vacation with a girlfriend in 1925. His wife stayed home. To aid his concentration, Schrödinger brought with him two pearls to keep noise out of his ears. Exactly what noise he wished to avoid is not clear. Nor do we know the identity of the girlfriend, nor whether she was inspiration or distraction. Schrödinger kept discreetly coded diaries, but the one for just this period is missing.

In four papers published within the next six months, Schrödinger laid down the basis of modern quantum mechanics with an equation describing waves of matter. Almost all the puzzles of the early quantum theory seemed resolved. The work was immediately recognized as a triumph. Einstein said it sprang from "true genius." Planck called it "epoch making." Schrödinger himself was delighted to think that he had gotten rid of quantum jumping. He wrote:

> It is hardly necessary to point out how much more gratifying it would be to conceive a quantum transition as an energy change from one vibrational mode to another than to regard it as a jumping of elec-

trons. The variation of vibrational modes may be treated as a process continuous in space and time and enduring as long as the emission process persists.

(The Schrödinger equation is actually a nonrelativistic approximation. That is, it holds only when speeds are not close to that of light. The conceptual issues we treat are still with us in the more general case. It is simpler, clearer, and also customary to deal with the quantum enigma in terms of the Schrödinger equation. And even though photons move at the speed of light, essentially everything we say applies equally to photons.)

History is more complicated than the story we just told, and more acrimonious. Almost simultaneously with Schrödinger's discovery, Bohr's young postdoc, Werner Heisenberg (of whom we'll hear more later), presented his own version of quantum mechanics. It was an abstract mathematical method for obtaining numerical results. It denied any pictorial description of what was going on. Schrödinger criticized Heisenberg's approach: "I was discouraged, if not repelled, by what appeared to me a rather difficult method of transcendental algebra, defying any visualization." Heisenberg was equally unimpressed by Schrödinger's wave picture. In a letter to a colleague: "The more I ponder the physical part of Schrödinger's theory, the more disgusting it appears to me."

For a while it seemed that two intrinsically different theories explained the same physical phenomena, a disturbing possibility that philosophers had long speculated about. But within a few months, Schrödinger proved that Heisenberg's theory was logically identical to his own, just a different mathematical representation. The more mathematically tractable Schrödinger version is generally used today.

The Wavefunction

Heisenberg did, however, have a point about the physical aspect of Schrödinger's theory. What's waving in Schrödinger's matter wave? The mathematical representation of the wave is called the "wavefunction." In some very real sense, the wavefunction of an object *is* the object. In quantum theory there is no atom in addition to the wavefunction of the atom. But what, exactly, is Schrödinger's wavefunction physically? At first, Schrödinger didn't know, and when he speculated, he was wrong. For now, let's just plow ahead and look at some wavefunctions that the equation tells us can exist. That's what Schrödinger did.

The essentials of quantum mechanics can be seen with the wavefunction of a simple little thing moving along in a straight line. It could be an electron or an atom, for example. To be general, we usually refer to an "object" but sometimes revert to "atom." We later discuss wavefunctions for bigger things—a molecule, a baseball, a cat, even the wavefunction of a friend. Cosmologists contemplate the wavefunction of the whole universe, and so will we.

A couple of years before Schrödinger's vacation inspiration, Compton showed that photons bounced off electrons as if they were each tiny billiard balls. On the other hand, to display interference, each and every photon or electron had to be a widely spread-out thing. Each photon, for example, had to go through both slits in a barrier. How can an object be both compact and spread out? Well, a wave can be *either* compact or spread out. (But, of course, it cannot be both at the same time.)

The wavefunction of a moving atom might look much like ripples, or a series of waves, a "wave packet," moving on water. A wave equation, the one for water waves or matter waves, can describe a spread-out packet with many crests, or a compact packet with only a few crests, or even a single crest moving along.

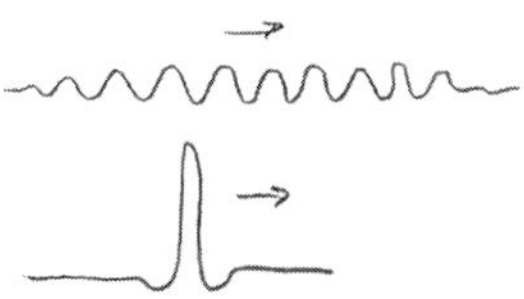

For big things, objects much larger than atoms, Schrödinger's equation just turns into Newton's universal equation of motion. Schrödinger's equation governs not only the behavior of electrons and atoms but also the behavior of everything made of atoms—molecules, baseballs, and planets. Given an initial wavefunction, it tells what the wavefunction will be like later. It's the *new* universal law of motion. Newton's equation is just the approximation for big things.

Figure 7.3 Wavefunction as a series of waves or a single crest

Waviness

Schrödinger's equation says a moving object is a moving packet of waves. But what's waving? Think of these analogies—Schrödinger no doubt did:

At a stormy place in the ocean, the waves are big. Let's call that a region of large "waviness." The boom of a drum, on its way to you from a distant drummer, is where the air pressure waviness is large, where the sound *is*. The bright patch where the sunlight hits the wall, the region of large electric field waviness, is where the light *is*. Waviness somehow tells where something *is*. It might seem reasonable to carry this notion over to the quantum case.

The waviness of a packet of quantum waves is large where the amplitude of the waves is large. Perhaps that is where the object *is*. (In quantum theory, the technical expression for the waviness is the "absolute square of the wavefunction," and there is a mathematical operation for getting it from the wavefunction. We mention that term only because you might see it elsewhere. "Waviness" is more

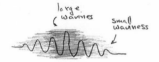

Figure 7.4 A wavefunction and
its waviness

descriptive.) Waviness can be easy to sketch if we have the wavefunction. We will indicate waviness by shading: The darker the shading, the greater the waviness.

When we considered an atom simply as a moving object, we ignored its internal structure. There are, of course, electron wavefunctions within the atom. Early on, Schrödinger calculated the wavefunction of the single electron within the hydrogen atom and duplicated Bohr's results for the experimentally observed hydrogen spectrum—without needing Bohr's arbitrary assumptions. Able to do that, Schrödinger was sure he had it right. He was elated. He thought he had gotten rid of quantum jumps. Not so, we'll see.

In figure 7.5, we sketch the waviness for the hydrogen electron's three lowest energy states as cross sections through the three-dimensional waviness. You can

visualize the waviness as clumps of fog: The fog is densest where the waviness is largest. Pictures such as these provide chemists with insight into how atoms and molecules bind with each other.

The wavefunction, being the object itself, actually includes everything knowable about an object, the velocity of an atom or its rate of spin, for example. For now, we will talk only

Figure 7.5 The waviness of a hydrogen atom's
three lowest states

of the wavefunction for the position of an object moving along a straight line. The quantum enigma confronts us starkly in that simple case.

We have suggested that the waviness perhaps tells where the object is. It's not quite that. But what exactly *is* the waviness?

Schrödinger's Initial (Wrong) Interpretation of Waviness

Schrödinger speculated that an object's waviness was the smeared out object itself. Where, for example, the electron fog is densest, the material of the electron is most concentrated. The electron itself would thus be smeared over the extent of its waviness. The waviness of one of the states of the hydrogen electron pictured above might then morph smoothly to another state without the quantum jumping Schrödinger detested.

This reasonable-seeming interpretation of waviness is wrong. Here's why: Though an object's waviness may be spread over a wide region, when one looks at a particular spot, one finds either a whole object or no object in that spot.

For example, an alpha particle emitted from a nucleus might have waviness extending over kilometers. But as soon as a Geiger counter detects an alpha, there is a whole alpha right there inside the counter. The waviness of a single electron having just passed through both slits in an interference experiment will be in several clumps, separated perhaps by inches, and each headed toward an allowed region on the screen. But an instant later a single flash is seen at a single spot on the scintillation screen, and the whole electron can be found there. All the electron's previously extended waviness is suddenly concentrated at that one spot at which the electron can be observed. If, on the other hand, the electron were observed in transit to the screen, it would be found somewhere in the clumps of waviness.

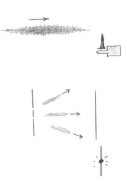

If an actual physical object were smeared over the extent of its waviness, its remote parts would have to instantaneously coalesce to the place where the whole object was found. Physical matter would have to move at speeds greater than that of light. That's impossible.

Schrödinger's equation succeeds in predicting what is actually seen, but in his goal of exorcising what he called the "nonsense" from physics, Schrödinger failed. He once claimed that if we must still put up with "these damn quantum jumps," he was sorry that he had anything to do with quantum theory. We later treat his objection to what quantum theory says about physical reality. It is something far more outrageous than mere orbit-jumping electrons.

Figure 7.6 *Top*: Waviness of an alpha particle before and after detection by a Geiger counter. *Bottom*: Waviness of a single electron before and after detection on a screen.

The Accepted Interpretation of Waviness

What we say in the next several pages makes this the most difficult chapter in the book—difficult because it's so hard to *believe*.

The waviness in a region is the probability of *finding* the object in that region. Be careful—the waviness is not the probability of the object *being* there. There's a crucial difference! The object was not there before you found it there. Your happening to find it there *caused* it to be there. This is tricky and the essence of the quantum enigma. Let's back up and try to see what this quantum probability might mean. We must contrast our usual understanding of probability with its role in quantum mechanics. Let's start with an example of classical probability.

At a carnival, a fast-talking fellow with even faster hands operates a shell game. He places a pea under one of two inverted shells. After his rapid shuffling, your eyes lose track of which shell holds the pea. There is equal prob-

Figure 7.7

ability for the pea to be in either of two places. We associate a probability of one-half with each shell, meaning half of the times we look we would find the pea under, say, the right-hand shell. (The sum of the probabilities for the two shells is one: $\frac{1}{2} + \frac{1}{2} = 1$. This corresponds to the certainty that the pea is surely under one of the two shells.)

After a bit of glib talk (as he takes some bets) the operator lifts, say, the shell on the right, and you see the pea. Instantaneously, it becomes a certainty (probability equal to one) that the pea is not under the shell on the left. The probability that the pea is under the left shell collapsed to zero. That left shell could have been moved across town before the shell on the right was lifted. The collapse of probability would still be instantaneous. Great distance does not affect how fast probability can change.

Figure 7.8

Games of chance make it almost obvious what waviness should represent. (Obvious at least to those of us who have been previously taught the answer.) It was, in fact, only a few months after Schrödinger announced his equation that Max Born realized that the waviness in a region was probability, the probability for the whole object being found in that region. Like probability in the shell game, when we find out where the object is, its waviness instantaneously becomes unity in the region we found it and zero everyplace else.

There is, however, a crucial difference between the classical probability illustrated by the shell game and that in quantum mechanics represented by waviness. Classical probability is a statement of one's knowledge. In the shell game, for instance, your not knowing at all which shell covered the pea means that for you the probability of it being under each shell was $\frac{1}{2}$. The shell game operator likely had better knowledge. If so, for him the probability was different.

Classical probability represents (someone's) knowledge of a situation. It does not tell the whole story. Something physical is presumed to exist in addition to that knowledge, something it was the probability *of*. There existed, for example, a real pea under one of the shells. If someone peeked and saw the pea under the left-hand shell, the probability would collapse to a certainty *for her*. But it could still be $\frac{1}{2}$ for each shell for her friend who didn't look. Classical probability is subjective.

Quantum probability, waviness, on the other hand, is objective—it's the same for everyone. It's the whole story: There is no atom in addition to the wavefunction of the atom. If someone happened to see the atom at a particular spot, that look would collapse the spread out wavefunction of the atom to be concentrated at that particular spot for everyone. Any subsequent looker would find the atom there—as long as they looked before it moved away.

(On top of the quantum probability there can, of course, also be ordinary uncertainty. The first looker's friend might not yet know that she had already collapsed the wavefunction. We can ignore this.)

In quantum theory there is no atom in addition to the wavefunction of the atom. This is so crucial that we say it again in other words: The atom's wavefunction and the atom are the same thing; "the wavefunction of the atom" is a synonym for "the atom." Accordingly, before a look collapses a widely spread-out wavefunction to the particular place where the atom is found, the atom did not exist there prior to the look. The look brought about the atom's existence at that particular place—for everyone.

We have been talking of an atom because quantum theory was developed to deal with microscopic objects. Later we'll just talk of "objects," because, in principle, quantum theory applies to everything and has been demonstrated for objects much larger than atoms.

At this point you may be mystified by quantum theory. (If so, you join many experts.) And it is now time for us to display the archetypal quantum experiment that demonstrates the creation of reality by observation: the so-called "two-slit experiment" treated in every quantum mechanics text. Our version is a bit like the shell game.

An Atom in a Box Pair

Our description is a simplified schematic of frequently performed demonstrations. In the shell game, the pea had equal probability for being under each shell. We will put equal parts of the waviness of a single atom simultaneously in each of two boxes.

Any wave can be reflected. Semitransparent mirrors reflect part of a wave and allow the rest to go through. A windowpane allows some light through and reflects some. A semitransparent mirror for light is a semitransparent mirror for photons. The wavefunction of each individual photon hitting a semitransparent mirror splits with part being reflected and part transmitted. We can also have semitransparent mirrors for atoms. Encountering such a mirror, an atom's wavefunction splits into two wave packets; one packet goes through, and another is reflected.

The arrangement of mirrors and boxes in figure 7.9 allows the trapping of the two parts of the wavefunction of an atom by closing the box doors when both

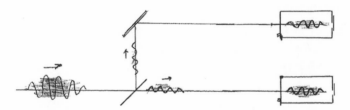

Figure 7.9 Mirror and box set-up allowing trapping of wavefunctions.
A wavefunction is shown at three different times.

packets are surely inside the boxes. We show the wavefunction and waviness at
three successive times.

Holding an atom in a box pair without disturbing its wavefunction would
be tricky, but possible. Dividing the wavefunction of an atom into two well-
separated regions is frequently accomplished, and that's all we really need for
our story. We like to think of each region defined by a box because it's more like
the shell game.

But, unlike the classical shell game, where the pea was in fact under one
shell or the other, quantum theory says the waviness, and therefore the atom, is
simultaneously in both boxes. What can that possibly mean? We establish that with
an interference experiment, the standard demonstration of the wave phenomena.
(Recall our description of interference in chapter 5.)

We open a small hole in each box of the pair at about the same time. The
wavefunction leaks out of *both* boxes and falls on a screen to which an atom will
stick. In some places on the screen, waves from the two boxes will reinforce each
other, and at other places waves will cancel. Repeating this with many identically
positioned box pairs, atoms will be found in regions of large waviness.

That's the crucial point: *Each and every* atom follows a rule allowing it to
land in regions separated by distance "d" in figure 7.10. That rule depends on the

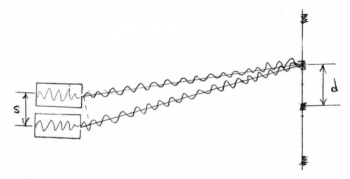

Figure 7.10 Interference experiment with two-box set-up

box-pair spacing "s." Therefore, each atom had to "know" the box-pair spacing. According to quantum theory, each atom knows the rule because each atom was in both boxes at the same time.

Wouldn't it make more sense to say that *part* of each atom was in each box? No, that doesn't work. How do we establish that?

Suppose, instead of doing an interference experiment, we just look in a box to see which one held the atom. It doesn't matter how we look. We can, for example, actually shine an appropriate light beam into the box and see a glint from the atom. About half the time we will find a whole atom in the looked-in box; about half the time we find the box empty. If there is no atom in the box we look in first, it will always be in the other.

But before we looked, an interference experiment could have established that the unobserved atom had been in both boxes. The atom didn't have a single position. But, on looking, we find the whole atom in a single box.

The most accurate way of describing the state of the unobserved atom is to put into English the mathematics describing the state of the atom before we looked to see where it is: The atom was *simultaneously in two states;* in the first state, it is in-the-top-box-and-not-in-the-bottom-box, and simultaneously in the second state, it is in-the-bottom-box-and-not-in-the-top-box.

Putting it this way, however, boggles the mind. It's saying a physical thing was in two places at the same time. The quantum mechanical term for this situation is that the atom is in a "superposition state" simultaneously in both boxes.

We should not leave this discussion without emphasizing that what we have said about the position of an object being created by observation applies to every other property. For example, an atomic nucleus is a tiny magnet with a north and a south pole. It can be in a superposition state with its north pole simultaneously pointing up and down.

An object in two places at once is so counterintuitive that it is inevitably confusing. Some of the confusion will probably be straightened out in our later chapters. But not all of it! We have confronted the still unresolved and definitely controversial quantum enigma. At this point, let us look at two attitudes about quantum probability.

Two Attitudes on Waviness as Probability

An "all's okay" attitude: Waviness is the probability of what you will *observe.* Yes, it depends on how you look. For our box pairs, it's whether you look in single boxes or whether you do an interference experiment and look at the screen after opening holes in both boxes. In either case, quantum theory predicts the correct result. Correct predictions are all one ever needs—for all practical purposes. We defend this useful pragmatic attitude, called the Copenhagen interpretation, in chapter 10.

A disturbed attitude: Does Nature's fundamental law, the Schrödinger equation, give only *probability*? Einstein felt that there must be an underlying deterministic explanation. "God does not play dice," is his often-quoted remark. (Bohr told him not to tell God how to run the universe.)

But randomness was not Einstein's most serious problem with quantum mechanics. What disturbed Einstein and Schrödinger, and more people today, is quantum mechanics' apparent denial of ordinary physical reality—or, maybe the same thing, the need to include the observer in the physical description—an intrusion of consciousness into the physical world.

In the shell game, the probability was that of a pea being under a particular shell. There was also an actual pea under one of the shells. According to quantum mechanics, there was *not* an actual atom in one of the boxes before we looked and found it there. But there *are* actual atoms, and actual things made of atoms. Aren't there?

If you're not a bit baffled at this point, you've missed the point. According to Richard Feynman, who understood quantum mechanics as well as anyone ever did: "Nobody understands quantum mechanics."

In our next short chapter we take an interlude to tell of practical things. Only then will we face up to physics' skeleton in the closet, its encounter with consciousness.

8

One-Third of Our Economy

Developing quantum theory was "the crowning
intellectual achievement of the last century," says
California Institute of Technology physicist John
Preskill. It's the underlying principle for many of
today's devices, from lasers to magnetic resonance
imaging machines. And these may prove to be just
the low-hanging fruit. Many scientists foresee
revolutionary technologies based on the truly strange
properties of the quantum world.

—*Business Week,* March 15, 2004

We were deep into the quantum mysteries in the fourth week of our "Quantum Enigma" course, which is addressed to students not majoring in the sciences (though some physics majors always take it), when a young woman's hand went up with a question: "Is quantum mechanics useful for anything *practical?*" I (Bruce) was speechless for at least ten seconds. In the narrowness of my physicist perspective, I just assumed that everyone realized the quantum basis of our technology. I put aside my lecture notes and for the rest of the hour went off on a tangent to tell of practical applications of quantum mechanics.

This short chapter takes us off on that same tangent. The theme of our book is presenting the undisputed quantum facts that reveal physics' encounter with consciousness. But the same quantum facts are basic to both modern science and today's technology. After the far-out stuff of our previous chapter, it's good to make contact with solid ground before taking off again.

Quantum mechanics is essential to every natural science. When chemists do more than follow empirical rules, their theories are fundamentally quantum mechanical. Why grass is green, what makes the sun shine, or how quarks behave inside protons are all questions that must be answered quantum mechanically. The still-to-be-understood nature of black holes or the Big Bang is sought in quantum terms. String theories that may hold the clue to such things all start with quantum mechanics.

Quantum mechanics is the most accurate theory in all of science. An extreme test is the calculation of the "gyromagnetic ratio of the electron" with a precision of a part in a trillion. (What the gyromagnetic ratio is doesn't matter here.) Measuring something that accurately is like measuring the distance from a point in New York to a point in San Francisco to better than the thickness of a human hair. But it was done, and the theory was right on the mark.

Quantum mechanics works well in science, but how important is it practically? In fact, one-third of our economy involves products based on quantum mechanics. Here we describe three technologies where the quantum aspects are right up front: the laser, the transistor, and MRI (magnetic resonance imaging). We won't go into detail—our point is to show how quantum phenomena enter the picture and how physicists and engineers deal with the seemingly contradictory properties of microscopic entities.

The Laser

Lasers come in a wide variety. Some are many meters long and weigh tons. Others extend less than a millimeter. The beam of red light scanning bar codes at the supermarket checkout counter comes from a laser. A laser reads DVDs and writes in laser printers. A powerful laser can drill through concrete. Lasers produce the light for fiber optic communication, set lines for surveyors, and guide "smart bombs." With a sharply focused laser, a surgeon can pin down a detached retina.

A laser produces a nondiverging beam of light of a single frequency that can be focused down to a tiny spot. The essential physical principle is "stimulated emission of radiation": When a photon of the proper frequency hits an atom in an excited state, it stimulates the emission of a second photon of exactly the same frequency traveling in exactly the same direction—a clone. Where we had one photon, we now have two identical photons. If we maintain many atoms in the excited state, this process continues in a chain reaction to produce many identical photons. "Laser" is an acronym for Light Amplification by Stimulated Emission of Radiation.

A problem the laser designer must surmount is that the likelihood of a photon hitting an atom on a single pass through the lasing material is small. The light is therefore passed back and forth through the material resonating between a pair of mirrors. A resonating guitar string must vibrate in an integral number of half wavelengths. Likewise, laser mirrors must be spaced an integral number of half wavelengths of the light. One of the mirrors is slightly transparent, allowing a bit of the beam to leave the laser on each bounce.

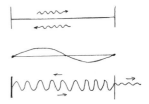

Figure 8.1 Light waves between laser mirrors

Notice how we slipped from talking of light being a stream of compact photons, each hitting a single atom, to light being an extended wave stretching between two macroscopic mirrors. (This is analogous to our atom that could be compactly concentrated in a single box or be a wave spread over two boxes.)

The Transistor

The transistor is the most important invention of the twentieth century. Without it, nothing dependent on modern electronics would be possible. The transistor can act as a switch, allowing an electric current to flow, or as an amplifier, taking in a weak electrical signal and putting out a stronger signal. Before the transistor was developed in the 1950s, such operations were done by vacuum tubes. Each tube was as large as your fist, gave off almost as much heat as a light bulb, and cost several dollars.

Today, a billion transistors on a single chip cost a millionth of a cent each, and each is only millionths of an inch across. A personal computer may have more than ten billion of them. Using vacuum tubes, a computer with the power of a modern laptop would be ridiculously expensive, occupy vast territory, and require all the electric power of a major city's generating plant.

Transistors are everywhere: in TVs, cars, cell phones, microwave ovens, and the watch on your wrist. Modern life depends on the transistor. In the year 2003, more than a hundred billion transistors were manufactured—every second.

Most transistors are based on silicon, each atom of which has fourteen electrons. Of these, four are "valence electrons" that bind each silicon atom to its neighbors. The other ten electrons are held to their parent nucleus, but each valence electron extends throughout the silicon crystal as a wave. Each valence electron is simultaneously everyplace in the crystal.

The electrons directly involved in the switching or amplifying functions of the transistor are another matter. These can be released by phosphorus atoms, which are added to the silicon crystal. Designers of transistors must concern themselves with these released "conduction electrons" being slowed by bumping into individual impurity atoms or being trapped by such impurities. They must treat conduction electrons as objects compact on the atomic scale.

How do the engineers and physicists who design lasers and transistors deal with photons and electrons that sometimes are smaller than atoms and sometimes extend over macroscopic distances? They cultivate a benign schizophrenia. They just learn when to think one way and when to think the other way. And, for all practical purposes, that's good enough.

Magnetic Resonance Imaging (MRI)

MRI produces strikingly clear and detailed images of any desired tissue in the body. It is on the way to becoming medicine's most important diagnostic tool. Presently, MRI machines are large and expensive, costing more than a million dollars. An MRI examination can cost well over a thousand. Fortunately, size and costs are coming down even as diagnostic capabilities increase.

Magnetic resonance images determine the distribution of a given element, usually hydrogen, in a particular material in the region of the body examined. Different tissues, bone or flesh, tumor or normal, are pictured by the differing concentrations of a particular chemical substance.

The details of MRI are complicated, but the only point we wish to make is that, as for the laser and transistor, physicists and engineers developing MRI must explicitly take quantum phenomena into account. The basic idea is the magnetic resonance of nuclei. (Magnetic resonance imaging was originally called "*nuclear magnetic resonance imaging*" before the anxiety-causing n-word was dropped.)

Nuclei are little magnets having a north and a south pole. In a magnetic field, the hydrogen nucleus, which is a proton, is "spatially quantized." That is, it has two energy states: In one, its north pole points up along the magnetic field; in the other, its north pole points down against the field. In an MRI machine, a radio wave of the proper frequency puts the hydrogen nuclei that are in the particular spot in the body being imaged at that instant into a quantum superposition state in which their north poles point up and down simultaneously. These nuclei emit radio waves as they return to their lower state, and the amount of this radio-frequency radiation reveals their concentration. A computer then creates the image.

Crucial to most MRI machines is a several-ton superconducting magnet held at a temperature a few degrees above absolute zero. In a superconducting metal, electrons condense into a quantum state in which they all move as a unit. Each electron is simultaneously everyplace in almost a ton of metal. Once the electrons are given an initial push, no electric power is needed to maintain the current flow and the magnetic field.

MRI is made possible by the coming together of the quantum phenomena responsible for nuclear magnetic resonance, superconductivity, and the transistor. Each of these technologies, as well as the laser, has led to a Nobel Prize in physics, MRI most recently, in 2004.

The Future

Quantum Dots The involvement of quantum mechanics in technology and biotechnology expands rapidly. In 2003, the journal *Science* named "quantum dot" research as one of the top scientific breakthroughs of the year. Quantum dots, each made of a few hundred atoms, are essentially artificial constructs with all the quantum properties of a single atom. Some have been designed to reveal the workings of the nervous system or to be ultrasensitive detectors of breast cancer. When electrodes are attached to quantum dots, they can be used to control current flow as ultrafast transistors or to process optical signals. Expect to hear a lot about quantum dots in the near future.

Quantum Computers An operating element in a classical digital computer must be in one of two states: either "0" or "1." An "unobserved" operating element in a *quantum* computer can be in a superposition state of simultaneously "0" and "1." This is much like the situation we described in our previous chapter, where a single unobserved atom was in a superposition state simultaneously in each of two boxes.

While each element in a classical computer can deal with only a single computation at a time, superposition allows each element in a quantum computer to deal with many computations simultaneously. This vast parallelism would enable a quantum computer to solve in minutes certain problems that would take a classical computer a billion years. Commercial applications are, however, not imminent. Quantum computers face some serious technical problems.

Engineers and physicists who work with the technologies we have spoken of may deal intimately with quantum mechanics on an everyday basis, but they never need to face up to the deeper issues raised by quantum mechanics. Many are not even aware of them. In teaching quantum mechanics, physicists, including us, minimize the enigmatic aspect in order not to distract students from the practical stuff they will need to use. We also avoid the enigma because it is a bit embarrassing; it's been called our "skeleton in the closet." In chapter 9 we look inside the closet.

9

Our Skeleton in the Closet

The interpretation [of quantum mechanics] has remained a source of conflict from its inception. . . . For many thoughtful physicists, it has remained a kind of "skeleton in the closet."

—J. M. Jauch

Just the facts, ma'am, just the facts.

—*Dragnet's* Sgt. Friday

In his book *Dreams of a Final Theory*, Nobel Laureate Steven Weinberg writes: "The one part of today's physics that seems to me likely to survive unchanged in a final theory is quantum mechanics." We share Weinberg's intuition about the ultimate correctness of quantum mechanics.

John Bell, a major figure in our later chapters who would likely have the Nobel Prize if it could be awarded posthumously, felt that "the quantum mechanical description will be superseded. . . . It carries in itself the seeds of its own destruction." Bell does not really disagree with Weinberg. His concern with quantum mechanics is not that an error will be found in any of its predictions, but that it is not the whole story. For him, quantum mechanics reveals the incompleteness of our worldview. He feels it is likely "that the new way of seeing things will involve an imaginative leap that will astonish us." (Incidentally, Bell tells that it was a lecture by Jauch—whom we quote just above—that inspired his investigations into the fundamentals of quantum mechanics.)

Along with Bell, we suspect that something beyond ordinary physics awaits discovery. Not all physicists would agree. Many would like to dismiss the enigma, our "skeleton in the closet," as merely a psychological problem, claiming that we just have to get used to the quantum strangeness.

However, the existence of an enigma is not a physics question. It's *meta-physics* in the original sense of that word. (*Metaphysics* is Aristotle's work that followed his scientific text *Physics*. It treats more general philosophical issues.) Here nonphysicists with a general understanding of the experimental *facts*—facts about which there is no dispute—can have an opinion with validity matching that of physicists.

We illustrate this point with a story in which an orthodox-minded physicist demonstrates some basic experimental facts of quantum mechanics to a group of rational and open-minded people (the GROPE) who have never been exposed to the quantum theory that explains those facts. What our physicist demonstrates to the GROPE is analogous to the experience of the visitor to Neg Ahne Poc. Though what was displayed in Neg Ahne Poc is *not* actually possible, that visitor's bafflement is the same bafflement the GROPE experiences from a demonstration that *is* actually possible. You may share that bafflement; we do—it's the quantum enigma.

After her demonstration, our physicist offers the standard quantum theory explanation for what was seen, the explanation that generally satisfies students in our quantum mechanics classes. Their concern with the physics calculations that will be on their exams overrides the interest in the meaning of what they calculate.

The "apparatus" our physicist uses is a caricature of an actual laboratory setup. But the quantum phenomena she demonstrates are well established for small objects. These phenomena are today being tried with ever-larger objects, now including midsized proteins. Is a virus next? Quantum theory sets no limit. The size of objects shown to exhibit such quantum effects seems constrained only by technology and budget.

We could be completely general in our story and talk of the experiments being done with "objects." That sounds vague. There's no reason we can't think of our objects as little green marbles. The experiment could actually be done with "little green marbles," as long as they were very little, say, the size of large molecules. So for our story, we talk in terms of marbles.

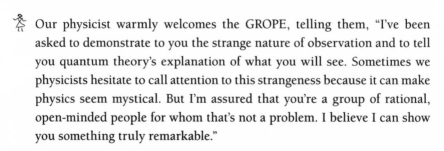 Our physicist warmly welcomes the GROPE, telling them, "I've been asked to demonstrate to you the strange nature of observation and to tell you quantum theory's explanation of what you will see. Sometimes we physicists hesitate to call attention to this strangeness because it can make physics seem mystical. But I'm assured that you're a group of rational, open-minded people for whom that's not a problem. I believe I can show you something truly remarkable."

The first experiment our physicist does should remind you of the visitor in Neg Ahne Poc asking: "In which hut is the couple?" The answers he got demonstrated the couple to be wholly in one hut or the other.

Our physicist points to a set of boxes, each box paired with another. She explains that with her apparatus she will inject a single marble into each *pair* of boxes. "The details of how my apparatus works," she says, "won't matter." The GROPE accepts this. They watch as she mounts a box pair on the right end of her apparatus, drops a tiny marble into a hopper on the left, and then removes the box pair. She repeats the procedure, accumulating a few dozen box pairs.

(Unlike the GROPE, you have *been exposed to quantum theory. We therefore note that our physicist's apparatus involves a set of mirrors appropriate for dividing the waviness of each "marble" equally into both boxes of each pair.)*

"My first experiment," our physicist explains, "will determine which box of each pair contains the marble." Pointing at a box pair, she nods to one eager-looking member of the GROPE and asks: "Would you please open each box and see which box holds the marble?"

Opening the first box, the young man announces: "Here it is."

"Make sure the other box is completely empty," requests our physicist.

Looking carefully, he says with assurance, "It's completely empty—there's nothing in it."

Once he was through examining the boxes, our physicist asks an attentive young woman to repeat the procedure of finding which box of a pair held the marble. Opening the first box, the woman remarks, "It's empty; the marble must be in the other box." Indeed, she finds it there.

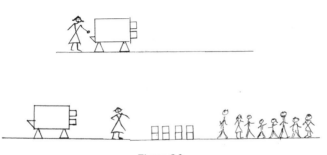

Figure 9.1

Our physicist repeats this procedure several more times. The marble appears randomly in either the first or second box opened. She soon notices members of the GROPE not paying much attention and mumbling to each other. She overhears one fellow say to the woman next to him: "What's her point? Hardly the remarkable demonstration we were promised."

Though the remark was not directed at her, our physicist responds: "I'm sorry, I just want to convince you that when we look to find out which box of a pair holds the marble, we demonstrate that there is a whole marble in one box and that the other is completely empty. Please bear with me, because I'd now like to show you that it doesn't matter just *how* we find out in which box our marble is. Here's another way to find out."

She sets a box pair in front of a sticky screen and opens one box. The light is too dim to see the fast marble, but there is a "plink," and a marble sticks to the screen. "Ah, the marble was in the first box," she says. "Therefore, no marble will hit the screen when I open the second box."

"Obviously," is a mumbled comment from someone near the back of the GROPE.

Though holding the attention of the GROPE again becomes difficult, our persistent physicist repeats the demonstration with more box pairs. If a marble hits the screen when she opens the first box, none appears when she opens the second box. If no marble appears on the screen on the first opening, there is always a marble on the second. The screen gradually becomes spotted with marbles, distributed uniformly over the screen.

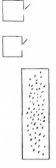

Figure 9.2
Opening boxes sequentially with results on the screen

"Can you see," she asks, "that this is also a demonstration that there is a marble in one of the boxes of a pair and that the other is empty?"

"Sure, but where's the *remarkable* demonstration you promised?" grumbles one fellow: "Of course *how* you look doesn't matter. Your apparatus put a marble in one box of each pair. So what?" Several nod agreement. And from an outspoken woman: "He's right!"

"Actually," our physicist says hesitantly, "the remarkable thing—what I hope to demonstrate—is that what he just said is not quite right. But let me try another experiment first."

The next experiment our physicist does should remind you of the visitor to Neg Ahne Poc asking: "In which hut is the man and in which hut is the woman?" The answers he got demonstrated the couple to be distributed over both huts.

The GROPE politely settles down to watch the new experiment.

Our physicist positions a new set of box pairs in front of the sticky screen and quickly opens both boxes of the pair. "The difference in this next experiment," she points out, "is that I'm opening both boxes at the same time." A plink indicates the impact of a marble on the screen. Discarding that box pair, our physicist carefully positions another in the same place and again opens both boxes together. Another plink is heard as a marble hits the screen.

Marbles accumulate on the screen as she opens more box pairs simultaneously. A fellow in a red shirt asks idly: "Doesn't this experiment demonstrate even less than your first one? Since you're now opening both boxes at the same time, for this set, we can't even tell which box the marble came out of."

But before his remark is seriously considered, a previously silent woman up front says: "Where the marbles land seems to form a pattern."

Now they all watch carefully. As more marbles plink onto the screen, the pattern emerges distinctly. Marbles land only in certain places. In other places on the screen there are no marbles. Each marble follows a rule allowing it to land only in certain places and forbidding it to land in others.

Figure 9.3
Results on screen of opening boxes simultaneously

The woman who first noticed the pattern seems puzzled and now asks: "In your first experiment, when the boxes of each pair were opened separately, the marbles were uniformly distributed over the screen. How can opening the empty box along with the one holding the marble affect where the marbles land?"

Our physicist, delighted with that question, responds eagerly: "You're right! Opening a box that was truly empty couldn't affect the marble. There was a marble in each pair of boxes. But it's not quite right to say that one box held the marble, and the other was empty. Each and every marble was simultaneously in both boxes of its box pair."

Responding to the dubious looks on the faces of most members of the GROPE, our physicist persists: "Actually, there's a quite convincing way to show that. It's just a bit time-consuming."

The GROPE chats and relaxes as our physicist quickly prepares three sets of box pairs. Now, regaining their attention, she repeats her simultaneous openings of both boxes of each pair. But this time with each of the three sets she uses a different spacing for the boxes of the pair.

Figure 9.4 Results of opening boxes simultaneously with different box spacings

 "Notice that the farther apart the boxes of a pair are, the closer spaced is the pattern. The rule that *each and every* marble obeys—the rule that tells each marble the places where it is allowed to land—depends on the spacing of its box pair. Each marble therefore 'knows' that spacing. Each marble must therefore have occupied both boxes of its pair."

"Wait a second, lady," a kid pipes up. "You're saying the marble was in *two places at the same time*, that it came out of both boxes. That's silly! . . . Ah, oh, I'm sorry, ma'am."

 "No problem, young fellow," responds our physicist. "You're quite right. The marble was simultaneously in two places. It was in both boxes. The scientific way is to accept what Nature tells us regardless of our intuitions. The single marble coming out of both boxes may sound silly, but the experimental demonstrations leave us no other alternative."

This takes a bit of contemplation. But after a minute or so, that fellow in the red shirt speaks up: "There is an alternative, an obvious one. In your first experiment, where you opened boxes one at a time, we saw one box of each pair to be completely empty. But, as you just said, for these other box pairs there was something in both boxes. Clearly these sets of box pairs were prepared differently."

 Our physicist pauses with her hands on her hips to allow this idea to take hold before she comments: "That's a reasonable hypothesis. But actu-

Figure 9.5 Drawing by Charles Addams.
© Tee and Charles Addams Foundation

Figure 9.6

ally the box pairs for both kinds of experiment were prepared identically. With either set I could have done either experiment. I'll prove that."

The third experiment our physicist does should remind you of the visitor to Neg Ahne Poc asking either question. He could choose to demonstrate either that the couple is in a single hut or that the couple is distributed over both huts. That baffled him.

After a coffee break, during which our physicist prepares and stacks up several sets of a dozen box pairs each, the GROPE reassembles. A woman speaks up: "We've been talking about what you said, and at least some of us are confused. A few of us think you claimed to demonstrate both that one box of each pair was empty and also that neither box was empty. Those are two contradictory situations. They misunderstood. Didn't they?"

"Well, they have it almost right. Which situation would you like to demonstrate with this group of box pairs?"

Somewhat taken aback, the questioner hesitates, but the woman next to her quickly volunteers: "Okay, show us that one box of each pair is empty."

Our physicist repeats her first experiment, opening the boxes of each of a dozen pairs in turn, each time revealing a marble in one of the boxes. Showing the other box to be empty, she comments: "And I assure you that no matter how this empty box is investigated, absolutely nothing would ever be found in it."

A cooperative fellow now points to another set of box pairs and asks: "Can you now show us that for this other set neither box is empty?"

"Sure, I'll do that." And our physicist performs the experiment that demonstrates each marble must have occupied both boxes of its pair, opening both boxes simultaneously for a dozen box pairs in a row.

Several times our physicist demonstrates either of the two apparently contradictory situations, as chosen by a GROPE member.

A fellow up front brusquely calls out in the middle of one of the demonstrations: "What you're telling us—and I admit seem to demonstrate—

makes no sense. It's logically inconsistent. . . . Oh, I apologize, I didn't mean to interrupt."

🧍 "No, no, it's okay," our physicist assures him. "You raise an important point."

He therefore continues: "You claim to demonstrate that both boxes of each pair contain at least parts of the marble, but you supposedly also show that one box of each pair is empty. That's logically inconsistent."

🧍 "You'd be right," replies our physicist, "if we showed both those results for the *same* set of box pairs. But since we actually did those two demonstrations with two different sets of marbles we see no logical inconsistency."

A woman objects: "But for the box pairs with which you demonstrated one thing, we *could* have asked you to demonstrate the opposite."

🧍 "But you didn't," is our physicist's almost too casual reply. "Predictions for not-done experiments can't be tested. Therefore, *logically,* there's no need for science to account for them."

"Oh, no, you can't squeeze through that loophole," the original objector retorts. "We're conscious human beings, we have free will. We could have made the other choice."

🧍 Our physicist squirms a bit: "Consciousness and free will are really issues for philosophy. Though I admit the issue of consciousness is raised by quantum mechanics, most of us, most physicists, prefer to avoid such discussion."

An earlier questioner is unsatisfied: "Okay," he demands, "but you agree that before we looked there was a matter of fact as to whether one box of each pair was empty or not. You physicists believe in a physically real world, don't you?"

He considered his question rhetorical. At least he expected a "Yes, of course" answer.

🧍 But our physicist hesitates, and again seems evasive: "What existed before we looked, what you call 'a physically real world,' is another issue most physicists prefer to leave to philosophers. For all practical purposes, all we need deal with is what we see when we actually do look."

"But you're saying something crazy about the world! You're saying that what previously existed is created by the way we look at something," is his unsatisfied response. Most heads nod in agreement; others just seem baffled.

"Hey, I promised I'd show you something remarkable. I've done that, haven't I?" Responding to some nods and some frowns, she continues: "We find the world stranger than we once imagined, perhaps stranger than we *can* imagine. But that's just the way it is."

"Wait!" a previously silent woman says firmly. "You can't get away with avoiding the issues your demonstrations raise. There's got to be an explanation. For example, instead of being in both boxes, maybe every marble has a kind of undetectable radar that tells it the separation of its box pair."

 "We can never rule out 'undetectable' things," our physicist admits. "But a theory with no testable consequences beyond merely those things it was invented to explain is unscientific. Just as useful as your theory of an 'undetectable radar' would be to assume that an invisible fairy rides on and guides each marble." Realizing she has embarrassed the proposer of the radar theory, our physicist apologizes: "I'm sorry; that was snide. Speculations like yours can be useful as jumping off points for developing testable theories."

"Oh, it's okay, I took no offense."

 "Actually, we already have a theory that explains everything I've demonstrated," continues our physicist, "and vastly more. It's quantum theory. It's basic to all of physics and chemistry, and much of modern technology. Even theories of the Big Bang are based on quantum theory."

"Why didn't you use it to explain your demonstrations?" questions a woman sitting with her chin in her hands.

 "I might have done that," replies our physicist, "but I wanted to make an important point: that the remarkable thing I've demonstrated, the quantum enigma—that the physical condition of the marble depends on your free choice of experiment—arises *directly from the experimental facts*. 'Just the facts, ma'am, just the facts,' as Sergeant Friday used to say. The quantum enigma is not merely *theoretical*. But now that you've seen the demonstration, let me tell you quantum theory's explanation of what we've seen."

"My apparatus," she continues, "puts a marble in each box pair, but it does not put that marble in a single one of the boxes. Let's talk about that very first experiment, the one in which you found a marble in one of the boxes and saw that the other box was empty.

"Quantum theory tells us that before you looked, the marble was in what we call a 'superposition state' simultaneously in both boxes. Your gaining knowledge of it being in a particular box *caused* it to be wholly in that box. Even if you gained that knowledge by finding one box empty and did not even see the marble, your merely gaining the knowledge that it was in the other box would cause it to be wholly in the other box. Gaining knowledge in any way whatsoever is enough."

The GROPE (being a group of reasonable and open-minded people) listens politely. But what our physicist said is not readily accepted.

A man suddenly blurts out: "Are you claiming that before we looked and found the marble in one of the boxes, it wasn't there, that our looking created the marble there? That'd be silly."

"Wait, I think I understand what she's saying," the woman sitting next to him volunteers. "I've read about quantum mechanics. I think she just means that the wavefunction, which is the probability of where the marble is, was in both boxes. The actual marble was, of course, in one box or the other."

"The first part of what you said is okay," says our physicist encouragingly. "What was in each of the boxes was indeed half of the marble's wavefunction. The waviness is the probability of finding a marble in the box. But there is no 'actual marble' in addition to the wavefunction of the marble. The wavefunction is the only thing that physics describes—it's the only physical thing."

Our physicist sees frowns and eyes rolled upward. She is glad they are (supposedly) open-minded. "Watch how nicely quantum theory explains the pattern we get when I open the boxes at the same time," she continues. "The parts of the wavefunction that were in each box reach the detecting screen together."

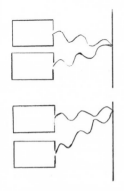

Figure 9.7
Reinforcement and cancellation of waves from two boxes

Moving her two hands wavelike as she talks: "The two parts of the wavefunction are waves moving out of each box to the screen. At some places on the screen, wave crests from one box arrive at the same time as crests from the other box, and the wavefunction from the two boxes add. That's a place with large waviness, with a large probability of finding a marble. At other places on the screen, crests from one box come at the same time as troughs from the other box, and the wavefunctions from the two boxes cancel each other. That's a place with no waviness, with zero probability of finding a marble. The distance of a place on the screen from each of the two boxes determines whether the wavefunctions from the two boxes will add or cancel. Each marble thus follows a rule forbidding it to appear in regions where the two parts of its wavefunction cancel. That explains the so-called pattern we see."

Satisfied that she has made her point, our physicist stands smiling with her hands on her hips.

A thoughtful-looking young woman slowly responds: "I understand how waves—you call them wavefunctions—can create patterns of waviness. We see that with light waves, even water waves. But probability has to be the probability of something. What's the waviness the probability of if it's not the probability of an actual marble being somewhere?"

"The wavefunction of the marble at some place gives the probability of your finding the marble there," our physicist emphasizes. "There was no actual marble there before you looked and found it there."

"I know this isn't easy to accept," she continues sympathetically. "Let me put it in other words. Consider a marble whose wavefunction is equally in both boxes. If you look in either box, you find out where the marble is. The probability becomes unity in one box and zero in the other. The waviness collapses totally into a single box. That concentrated waviness, which your observation *created*, is what you call the actual marble. But our being able to see an interference pattern proves there was no actual marble in a single box, before you looked."

"Just wait a second!" says a fellow who's been frowning and shaking his head for some time. "You put it in other words, so I'll put it in other words: If quantum theory says that by observing something someplace, I create it there, it's saying something ridiculous!"

"Would you say, 'shocking,' perhaps?" replies our physicist. "Niels Bohr, a founder of quantum theory, once said that anyone not shocked by quantum mechanics has not understood it. But no prediction of the theory has ever been wrong. You agree, don't you, that making consistently correct predictions is the *only* criterion a scientific theory need satisfy? That's been the method of science since Galileo."

At this point, another member of the GROPE can no longer contain herself: "If you're saying that unobserved things are just probabilities, that nothing's real until we observe it, you're saying we live in a dream world. You're trying to foist some silly solipsism on us."

"Well," our physicist replies calmly, "there's a saving grace. The big things we actually deal with are real enough. Remember, you need to do an interference-type experiment to actually demonstrate the creation by observation. And it's not practical—at least not yet—to do that with big things. So, for all practical purposes, there's no need for concern."

While that disturbed member fumes silently, another raises a hesitant hand and says: "If little things are not real, how can big things be real? After all, big things are just collections of little things. A water molecule is just one atom of oxygen and two of hydrogen, and an ice cube is just a collection of water molecules, and a glacier is just a big ice cube. Do we create the glacier by looking at it?"

Our physicist is now visibly uncomfortable. "Well, in a sense . . . , it's sort of complicated . . . but, as I said, for all practical purposes it doesn't really matter, so. . . ." Then noticing a member of the GROPE with a friendly expression, our physicist invites his comment with a smile.

Trying to be conciliatory, he volunteers: "Maybe what you're driving at is the notion that 'We create our own reality.' I sometimes feel much that way."

"Oh, I can go along with that," our physicist nods. "But that kind of 'reality' is something different. When I say, 'I create my own reality,' I'm talking of *subjective* reality. I'm saying that I accept responsibility for my personal perceptions and my social situation—something like that at least. The reality we're talking about here is objective reality, physical reality. An observation creates an objective situation, which is the same for everyone. After your looking in one of the boxes collapses the wavefunction of the marble into a particular box, anyone else who looks will find the marble there."

That member of the GROPE who had been fuming in silence now speaks up a bit too loudly: "This reality creation you're talking about is crazy! Your quantum theory may *work* perfectly, but it's absurd! Are people letting you physicists get away with this?"

"I suppose so," replied the physicist.

"Then you're getting away with murder!"

"Well, we usually keep the skeleton in the closet."

10

Wonderful, Wonderful Copenhagen

Wonderful, wonderful Copenhagen . . .

Salty old queen of the sea

Once I sailed away

But I'm home today

Singing Copenhagen, wonderful, wonderful

Copenhagen for me.

—"Wonderful Copenhagen," by Frank Loesser

The meaning of Newton's mechanics was clear. It described a reasonable world, a "clockwork universe." It needed no "interpretation." Einstein's relativity is surely counterintuitive, but no one interprets relativity. We get used to the idea that moving clocks run slow. It's harder to accept that observation *creates* the reality observed. That needs interpretation.

Students come into physics to study the down-to-earth physical world. The *Oxford English Dictionary* defines this sense of "physical" well: "Of or pertaining to material nature, *as opposed to the psychical, mental, or spiritual*" (emphasis added). The *New York Times* recently quoted science historian Jed Buchwald: "Physicists . . . have long had a special loathing for admitting questions with the slightest emotional content into their professional work." Indeed, most physicists want to avoid dealing with that skeleton in our closet, the role of the conscious observer. The Copenhagen interpretation of quantum mechanics allows that avoidance. It is our discipline's "orthodox" position.

The Copenhagen Interpretation

Niels Bohr recognized early on that physics had encountered the observer and that the issue had to be addressed:

> The discovery of the quantum of action shows us, in fact, not only the natural limitation of classical physics, but by throwing a new light upon the old philosophical problem of the *objective existence of phenomena independently of our observations, confronts us with a situation hitherto unknown in natural science.* (emphasis added)

Within a year after Schrödinger's equation, the Copenhagen interpretation was developed at Bohr's institute in Copenhagen with Niels Bohr as its principal architect. Werner Heisenberg, his younger colleague, was the other major contributor. There is no "official" Copenhagen interpretation. But every version grabs the bull by the horns and asserts that *an observation produces the property observed.* The tricky word here is "observation."

Copenhagen softens this assertion by defining an observation as taking place whenever a microscopic, atomic-scale, object interacts with a macroscopic, large-scale object. When a piece of photographic film is hit by a photon and records where the photon landed, the film has "observed" the photon. When a Geiger counter clicks in response to an electron entering its discharge tube, the counter has observed the electron.

The Copenhagen interpretation considers two realms: there is the macroscopic, classical realm of our measuring instruments governed by Newton's laws; and there is the microscopic, quantum realm of atoms and other small things governed by the Schrödinger equation. It argues that we never deal *directly* with the quantum objects of the microscopic realm. We therefore need not worry about their physical reality, or their lack of it. An "existence" that allows the calculation of their effects on our macroscopic instruments is enough for us to consider. Since the difference in scale between atoms and Geiger counters is so vast, it's okay to treat the microscopic and macroscopic realms separately.

Actually, in 1932, just a few years after Bohr's Copenhagen interpretation, John von Neumann presented a rigorous treatment also referred to as the Copenhagen interpretation. He showed that if quantum mechanics applies universally—as claimed—an ultimate encounter with consciousness is inevitable. Accordingly, Bohr's separation of the microscopic and the macroscopic is only a very good approximation. We discuss von Neumann's conclusion further in chapter 16. But keep in mind that whenever we refer to "observation," the question of consciousness lurks.

Figure 10.1 Drawing by Michael Ramus, 1991.
© American Institute of Physics

Physicists wishing not to confront philosophical problems readily accepted Bohr's version of the Copenhagen interpretation. Some physicists occasionally sail away to speculative shores. But when we actually *do* physics or teach physics, we all come home to wonderful Copenhagen.

Some physicists are concerned about denying reality to atoms and yet blithely viewing objects made of atoms as real. This is especially true as today's technology increasingly moves into the ill-defined region between the classical and the quantum realms. We therefore carefully examine the Copenhagen interpretation, the tacitly accepted, but increasingly questioned, stance of working physicists.

What Copenhagen Must Make Acceptable

While we presented the "skeleton in the closet" in our previous chapter as a story, those experiments, and many more like them, are done all the time—even as lecture demonstrations. The story was a caricature of an actual quantum experiment in which a small object is sent to occupy a pair of well-separated boxes. Looking into the boxes, you will always find the whole object in a single box, and the other box will be empty.

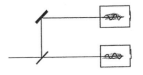

Figure 10.2 The atom-in-a -box -pair demonstration

According to quantum theory, however, before it was observed, the object was simultaneously in *both* boxes. And you could have chosen to do an interference experiment establishing that fact. Thus, by your free choice, you could establish either of two contradictory prior realities. And, in principle, quantum mechanics applies to everything—to baseballs as well as atoms. It's just our present technology that limits us to displaying quantum phenomena only with small things. That physical reality depends on our observation of it is what Copenhagen tries to make acceptable.

Since technology is not a limit in principle, we started our exploration of the quantum enigma with a bit of fantasy. We told of a visitor to Neg Ahne Poc, a place without our technological limitations. There he could experience something like quantum phenomena with large objects. When he asked in which hut is the couple, he was shown the couple together in a single hut. When he asked in which hut is the man and in which the woman, he was shown the couple separated; the couple occupied both huts. The couple's prior reality depended on the question he asked, the "experiment" he did. That baffled him. The explanation the Rhob offered was essentially the Copenhagen interpretation. (Neg Ahne Poc is Copenhagen spelled backward.)

Three Pillars of Copenhagen

The Copenhagen interpretation rests on three basic ideas: the probability interpretation of the wavefunction, the Heisenberg uncertainty principle, and complementarity. We look at these in turn.

The Probability Interpretation
of the Wavefunction

We've been using the idea all along that the waviness in a region (technically, the absolute square of the wavefunction) is the probability that the object will be found in that region. This probability interpretation of waviness is central to the Copenhagen interpretation.

While classical physics is strictly deterministic, quantum mechanics tells of an ultimate randomness in Nature. On the atomic level, God plays dice, Einstein notwithstanding. That Nature is ultimately statistical is not too hard for most people to accept. After all, much of what happens in everyday life has randomness. Were that the whole story, there would be little concern with a "quantum enigma." Probability in quantum mechanics implies something far more profound than randomness.

Classical probability in the shell game, say, is the *subjective* probability (for you) of where the pea is. In addition to that expression of your knowledge or uncertainty, there is a real pea under one shell or the other. Quantum probability is *not* the probability of where the atom is. It's the objective probability of where you (or anyone) will find it. The atom wasn't in that box before you observed it to be there. Quantum theory has the atom's wavefunction occupying both boxes. Since the wavefunction is synonymous with the atom itself, the atom is simultaneously in both boxes.

The point of that last paragraph is hard to accept. That's why we keep repeating it. Even students completing a course in quantum mechanics, when asked what the wavefunction tells, often incorrectly respond that it gives the probability of where the object *is*. The text we teach from emphasizes the correct point by quoting Pascual Jordan, one of the founders of quantum theory: "Observations not only *disturb* what is to be measured, they *produce* it." But we're sympathetic with our students. Using quantum mechanics is hard enough without worrying about what it means.

Though we've been speaking of "observation," we've not fully said what constitutes an observation. When a photon bounces off an isolated atom, does that photon observe the atom? Does a piece of photographic film hit by a photon observe that photon?

For a photon bouncing off an atom, there is a clear answer: The photon does *not* observe the atom. After the encounter, the photon is a wave of probability moving off in all directions. The photon and atom are in a superposition state that includes all possible positions of the atom before their encounter. This can be confirmed with a complex two-body interference experiment. According to Copenhagen, only when a macroscopic measuring instrument records the direction along which the photon came away from the atom does the existence of the atom in a particular position become a reality.

More generally, Copenhagen assumes that whenever any property of a microscopic object affects a macroscopic object, that property is "observed" and becomes a physical reality.

Strictly speaking, of course, a macroscopic object must still obey quantum mechanics and—if isolated from the rest of the world—merely joins the superposition state of the

Figure 10.3 Bouncing a photon off an atom does not create the atom's position until the photon is detected

microscopic object that affected it. It would thus not "observe." But for practical reasons, it is not possible to demonstrate that a large object is in a superposition state.

Though we have talked only of an object's position, in the Copenhagen interpretation *no* property of a microscopic object exists until it is produced by observation. Since an object is nothing but the sum of its properties (what else is there?), some argue that a totally unobserved microscopic object has no physical existence at all.

Let us be more careful about what is "unobserved." Consider our atom in its box pair. Until the position of the atom in a particular box is observed, the atom doesn't exist in a particular box. We nevertheless initially "observed" the atom when we grabbed it and sent it into our box-pair apparatus. The atom's position in the pair of boxes is thus an observed reality. However, taking the extreme case of very large boxes, we can simply say the atom has no position at all. It does not have the property of position. The same argument can be given for any other property of an object.

The Copenhagen interpretation generally adopts the simple view that only the *observed* properties of microscopic objects exist. Cosmologist John Wheeler puts it concisely: "No microscopic property is a property until it is an observed property."

If we carry this to its logical conclusion, microscopic objects themselves are not real things. Here's Heisenberg on this:

> In the experiments about atomic events we have to do with things and facts, the phenomena that are just as real as any phenomena in daily life. *But the atoms or elementary particles themselves are not real;* they form a world of potentialities or possibilities rather than one of things or facts. (emphasis added)

According to this view, atomic-scale objects exist only in some abstract realm, not in the physical world. If so, it's okay that they don't "make sense." It's enough that they affect our measuring instruments in accord with quantum theory. Those big things do "make sense," and we can consider them physically real and treat them with classical physics. But, of course, that classical description of their behavior is only an *approximation* to the correct quantum laws of physics. If so, in some sense, the microscopic realm, the unobserved realm, is the more real. Plato would like that.

However, if the microscopic realm consists merely of possibilities, how does physics account for the the small things that big things are made of? The most famous statement on this is often attributed to Bohr:

> There is no quantum world. There is only an abstract quantum description. It is wrong to think that the task of physics is to find out how nature *is*. Physics concerns what we can *say* about nature. (emphasis added)

This is actually a summary of Bohr's thinking by one of his associates. But it fits with what Bohr has said in more complicated ways. The Copenhagen interpretation avoids involving physics with the conscious observer by redefining what has been the goal of science since ancient Greece: to explain how the world actually works.

Einstein rejected Bohr's attitude as defeatist, saying he came to physics to discover what's really going on, to learn "God's thoughts." Schrödinger rejected the Copenhagen interpretation on the broadest grounds:

> Bohr's standpoint, that a space-time description [where an object *is* at some time] is impossible, I reject at the outset. Physics does not consist only of atomic research, science does not consist only of physics, and life does not consist only of science. The aim of atomic research is to fit our empirical knowledge concerning it into our other thinking. All of this thinking, so far as it concerns the outer world, is active in space and time. If it cannot be fitted into space and time, then it fails in its whole aim, and one does not know what purpose it really serves.

Would Bohr actually deny that a goal of science is to explain the natural world? Perhaps not. He once said: "The opposite of a correct statement is an incorrect statement, but the opposite of a great truth may be another great truth." Bohr's thinking is notoriously hard to pin down.

A colleague of Heisenberg's once suggested that the wave–particle problem is merely semantic and could be solved by calling electrons neither waves nor particles but "wavicles." Heisenberg, insisting that the philosophical issues raised by quantum mechanics included the big as well as the small, replied:

> No, that solution is a bit too simple for me. After all, we are not dealing with a special property of electrons, but with a property of all matter and of all radiation. Whether we take electrons, light quanta, benzol molecules, *or stones,* we shall always come up against these two characteristics, the corpuscular and the undular. (emphasis added)

He's telling us that, in principle (and that is what's important to us here), *everything* is quantum mechanical and ultimately subject to the enigma. This brings us to the second pillar of the Copenhagen interpretation, the uncertainty principle, the idea for which Heisenberg is most widely known.

The Heisenberg Uncertainty Principle

Heisenberg showed that any demonstration to refute the Copenhagen interpretation's claim of observer-created reality would be frustrated. Here's his example:

While doing an interference experiment, look to see out of which box each atom actually came. Seeing that would demonstrate that the atom had actually been in a single box, in spite of its following the rule implying that it came out of both boxes. Quantum theory (or at least its Copenhagen interpretation) would thus be shown inconsistent, therefore wrong. To show that any such demonstration must fail, Heisenberg produced the thought experiment that is now called the "Heisenberg microscope."

To see out of which box an atom came, you could bounce light off it—this is the usual way of seeing things. In order not to kick the atom hard enough to deflect it from an allowed place in the interference pattern, hit it with the least possible light, a single photon. To tell which box the atom came from, the wavelength of the light must be smaller than the separation of the boxes.

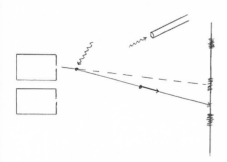

Figure 10.4 The Heisenberg microscope

But such a required short wavelength means a large number of crests coming per second. That's a high frequency, and a high-frequency photon has a high energy. It would give the atom a hard kick. Heisenberg easily calculated that photons with short enough wavelength would kick atoms hard enough to smear any interference pattern. Thus, if you saw each atom come from a single box, you could not *also* see an interference pattern showing that each atom had been in both boxes.

Werner Heisenberg proudly came to Bohr with his discovery. Bohr was impressed, but told his young colleague that he didn't have it quite right. Heisenberg forgot that if you knew the angle at which the photon bounced off, you *could* in fact calculate which box the atom came from. He had the right basic idea, though. Bohr showed him that by including the microscope needed to measure the photon angle in his analysis, he could recapture the result he thought he had. Missing this point doubly embarrassed Heisenberg. He reported that determining the direction of a light wave with a microscope was a question he had missed on his doctoral exam.

Heisenberg went on to generalize his microscope story to become the "Heisenberg uncertainty principle": The more accurately you measure an object's position, the more uncertain you will be about its speed. And vice versa, the more accurately you measure an object's speed, the more uncertain you will be about its position.

The uncertainty principle can also be derived directly from the Schrödinger equation. In fact, the observation of any property makes a "complementary" quantity uncertain. Position and speed are, for example, complementary quantities. Energy and the time of observation are another complementary pair. The bottom line is that any observation disturbs things enough to prevent the disproof of quantum theory's assertion that observation creates the property observed.

This is a good place for an example of why you can't display quantum strangeness with large objects. To see interference, an object's waves must be spread by passing through an opening smaller than its wavelength. Even a slow-moving grain of sand would have enough energy (momentum, actually) to make its wavelength smaller than the sand grain itself. But if the grain is larger than its wavelength, and the opening is smaller than that wavelength, the grain could not go through the hole. Actually, interference *would* be possible if the grain of sand moved slowly enough. But at the slowness needed, it would move less than an atom's length in a century. We're not patient enough for such experiments.

We shouldn't leave our discussion of the uncertainty principle without noting that it comes up in discussions of free will. In classical physics, if "an all-seeing eye" knew the position and velocity of every object in the universe at one moment, the entire future could be predicted with certainty. To the extent we are just part of this physical universe, classical physics rules out free will. Because the uncertainty principle denies this Newtonian determinism, it has stimulated heated philosophical discussions of determinism and free will. Quantum randomness is, of course, not free choice. The issue will be with us in our later chapters.

The uncertainty principle isn't quite enough to protect quantum theory from a seeming contradiction. We need Copenhagen's third pillar: complementarity.

Complementarity

Copenhagen invokes the "complementarity" principle to confront a spooky aspect of observation: the instantaneous collapse of an object's wavefunction *everywhere* by an observation *anywhere*.

Consider this experiment: With a set of box pairs, each pair containing an atom in a superposition state simultaneously in both boxes, look in one box of each pair. About half the time you will see an atom in the box you opened. According to the uncertainty principle, seeing that atom disturbed it with the photons you shined in. Therefore, toss away all the box pairs for which you saw and thus disturbed an atom. Merely using the uncertainty principle, you are left with a subset of box pairs whose atoms were *not* physically disturbed; no photons bounced off them. But for these box pairs, you know which box each atom is in: the box you did not look in.

Treat this subset of box pairs like any other set of box pairs and attempt an interference experiment. An interference pattern would prove that each of those atoms had been simultaneously in both boxes of its pair. But for this subset of box pairs, you already determined that each atom was wholly in a single box, the one you did not open. Finding an interference pattern would show an inconsistency in quantum theory.

In fact, these supposedly undisturbed atoms do *not* produce an interference pattern. These atoms don't follow the rule they would have followed had you not looked in the boxes that turned out to be empty. What caused these presumably undisturbed atoms to adopt a different behavior? After all, if you had done an interference experiment with those same atoms before you looked into the empty boxes, they would have produced an interference pattern.

Although these atoms were not physically disturbed—they did not deflect any photons—you did find out which box each atom was in. Your mere acquisition of that knowledge was sufficient to physically concentrate each atom totally within a single box. To avoid seeing this as somehow mysterious requires some talk.

The talk we offer in a quantum mechanics class for physics students is that when we look in a box and find no atom, we instantaneously collapse the atom's waviness into the other box. In the shell game, our look collapses the probability, which had been ½ for the pea being in each box, to being zero in the box we found empty and to 1, certainty, for the pea being in the other box. Essentially, the same thing happened with the waviness. After all, waviness *is* probability.

That's a bit glib. Classical probability *starts out* as a measure of one's knowledge. On the other hand, quantum probability, the waviness, is all there is to the physical atom. But we rarely emphasize philosophical conundrums to students, who must mainly learn to calculate.

Niels Bohr realized that he had to confront the spooky connection of knowledge with physical phenomena in order to allow physicists to just get on with doing physics without becoming involved with philosophy. He arbitrarily asserted his principle of complementarity: The two aspects of a microscopic object, its particle aspect and its wave aspect, are "complementary," and a complete description requires both contradictory aspects, *but we must consider only one aspect at a time.*

We avoid the seeming contradiction by considering the microscopic system, the atom, not to exist in and of itself. We must always include in our discussion—implicitly at least—the different macroscopic experimental apparatuses used to display each of the two complementary aspects. All is then fine, because *it is ultimately only the classical behavior of such apparatus that we report.* In Bohr's words:

> The decisive point is to recognize that the description of the experimental arrangement and the recording of observations must be given in plain language, suitably refined by the usual physical terminology. This is a simple logical demand, since by the word "experiment" we can only mean a procedure regarding which we are able to communicate to others what we have done and what we have learnt.

> In actual experimental arrangements, the fulfillment of such re-
> quirements is secured by the use, as measuring instruments, of rigid
> bodies sufficiently heavy to allow a completely classical account of
> their relative positions and velocities.

In other words, although physicists talk of atoms and other microscopic entities as if they were actual physical things, they are really only concepts we use to describe the behavior of our measuring instruments. They are not real, independent things like peas or stones, which we can speak about directly. Oh, yes, peas and stones are, strictly speaking, quantum mechanical. But, no matter, for all practical purposes big things allow a classical description. And, according to Copenhagen, that's all we need concern ourselves with.

This stance recalls Newton's *hypotheses non fingo* ("I make no hypotheses"), his claim that an explanation of gravity need not go beyond his equations predicting the motions of the planets. Einstein, of course, with general relativity, his theory of gravity, gave great insight into the nature of space and time by going beyond Newton's equations. May we one day go beyond complementarity?

Here's a slightly different tack that the flexible Copenhagen interpretation can take to avoid worrying about observer-created reality. It's in the spirit of complementarity. It asserts that it is meaningless to discuss experiments that might have been done but were in fact not done. After all, if you do an interference experiment demonstrating each object to have been simultaneously in both boxes, you could not then show *those same objects* to have been wholly in a single box.

The quantum enigma arises, afterall, through the presumption that you could have done a different experiment with the same objects. If we deny the need to account for observations that could have been made, but in fact were not made, we see no problem. We can just assume that with those objects that were in fact wholly in a single box, that's what we chose to demonstrate. With those objects that were in fact simultaneously in both boxes, that's what we chose to demonstrate. Our choices were correlated with what was in the box pairs. They were not truly free choices.

This situation is indistinguishable from a completely deterministic world. In a sense, it's worse—it's a conspiratorial world. Not only were our choices not free, but the universe conspired to correlate them with the different natures of the objects that were in the box pairs. In any event, taking this tack, the Copenhagen interpretation seems to deny free will.

Most of us can't accept that denial. We (Fred and Bruce) are each sure of our own free will—even though neither of us can be absolutely sure that his coauthor is not a sophisticated robot.

Our discussion of quantum theory and its Copenhagen interpretation dealing with hard-to-grasp notions such as wavefunction collapse, reality creation, free will, and consciousness might leave any reasonable person pretty far up in the air. So let's sidestep quantum theory and its interpretation for a moment and come to ground by restating the basic experimental facts from a theory-neutral point of view.

In the box-pair demonstration, you can look and see that each object is wholly in a single box. But you could have done an interference experiment, and you could have shown that the object was not wholly in a single box. You can therefore conclude that your look brought about the reality of the object being wholly in the box in which you found it. Moreover, by that conscious, free choice of which experiment to do, you could demonstrate either of two contradictory prior situations: a concentrated object or a spread-out object. Thus restating the demonstrable facts we might avoid some spooky theory; but, of course, you still simply have to accept the bafflement.

The Acceptance of—and the
Unease with—Copenhagen

The Copenhagen interpretation asks us to accept quantum mechanics pragmatically. (Bumper-sticker summary of pragmatism: "If it works, it's true.")

When physicists want to avoid dealing with philosophy, and for most of us that's almost all the time, we tacitly accept the Copenhagen interpretation. Physicists tend to be pragmatists. In dealing with microscopic objects, we analyze and report on the behavior of our laboratory apparatus. These big things present no paradox; they never need be considered to be in superposition states.

The properties of microscopic objects are *inferred* from the behavior of our apparatus. Nevertheless, we talk of microscopic objects, visualize them, and calculate with models of them as if they were as real as little green marbles. But if confronted with paradox, we retreat to the Copenhagen interpretation that microscopic objects are just theories. They should accurately explain the sensible behavior of our macroscopic equipment, but microscopic objects themselves need not "make sense."

Consider an analogy from psychology (as Bohr did). Basically, we report on and analyze a person's behavior. The physical behavior itself presents no paradox. A person's motives, however, are theories that should accurately predict the person's behavior, but motives need not, and often do not, make sense. We pragmatically accept this stance in dealing with people. The Copenhagen interpretation asks us to accept this stance in dealing with microscopic physical phenomena.

Bohr and others gave the Copenhagen interpretation broad philosophical underpinnings. But even when just accepted at face value, it provides a logical basis for physicists to get on with the practical aspects of physics without concerning themselves with the deeper meanings.

If you're not put at ease with Copenhagen's solution to the observer problem, you're not alone. When the two of us think honestly about what's really going on, we're always a bit bewildered. And we don't know of anyone who understands and takes seriously what quantum mechanics seems to be telling us who doesn't also admit some bafflement.

Nevertheless, until recently, most quantum mechanics textbooks implied that Copenhagen resolved all problems. One 1980 text dismissed the enigma with a joke, a sketch of a duckbilled platypus labeled "The classical analog of the electron." The idea was that going to the realm of the small, you should be no more surprised by an object being both an extended wave and a compact particle than zoologists going to Australia were surprised by an animal being both a mammal and an egg-laying "duck." In his preface, another 1980s author promises to "make quantum mechanics less mysterious for the student." He does it by never displaying the mystery.

Such attitudes likely stimulated Murray Gell-Mann's remark in his lecture accepting the 1976 Nobel Prize: "Niels Bohr brainwashed a whole generation of physicists into believing the problem had been solved." Gell-Mann's concern is a bit less relevant today since most current quantum texts at least hint of unresolved issues.

Essential to the Copenhagen interpretation was a clear separation of the quantum microworld from the classical macroworld. That separation depended on a vast difference in scale between atoms and the things we deal with directly. In Bohr's day, there was a wide no-man's land in between. It seemed acceptable to think of the macro realm obeying classical physics and the micro realm obeying quantum physics.

Today's technology has invaded the no-man's land. With appropriate laser light we can see individual atoms with the naked eye the way we see dust motes in a sunbeam. The "macroscopic apparatus" in this case is the human eye. With the scanning tunneling microscope (STM) not only can we see individual atoms, but we can pick them up and put them down. Physicists have spelled out their company's name by positioning thirty-five argon atoms. Atoms can now seem as real as little green marbles.

Figure 10.5 Thirty-five argon atoms.
Courtesy IBM

Quantum mechanics is increasingly applied to larger and larger objects. Even a one-ton bar proposed to detect gravity waves must be analyzed quantum mechanically. In cosmology, a wavefunction for the whole universe is written to study the Big Bang. It gets harder today to nonchalantly accept the realm in which the quantum rules apply as somehow not being physically real.

Nevertheless, many physicists pressed to respond to the strange nature of the microworld might say something like: "That's just the way Nature *is*. Reality is just not what we'd intuitively think it to be. Quantum mechanics forces us to abandon naive realism." And leave it at that.

Everyone is willing to abandon naive realism. But few of our colleagues are willing to abandon "scientific realism," defined as "the thesis that the objects of scientific knowledge exist and act independently of the knowledge of them." Admitting that quantum theory says that the existence of objects of the microworld depends on the knowledge of them, they would claim that the "knowledge" held by, say, a Geiger counter is sufficient to bring about that existence. (We deal with this in chapter 16, where the mystery of consciousness meets the enigma of quantum mechanics.)

Most physicists don't want to talk much about the implications of quantum mechanics. If pressed, few deny a quantum mystery but contend that the Copenhagen interpretation, or its modern extension, "decoherence" (discussed in chapter 14), has taken care of it—for all practical purposes at least; and that's all that counts.

But more physicists, especially younger physicists, are increasingly open-minded about ideas beyond Copenhagen. Other, wilder, interpretations proliferate, and we later discuss them. In recent years, concern with consciousness itself (as well as its connection with quantum mechanics) has emerged among philosophers and psychologists—even among neurologists. How come? One explanation offered is that the "mind-expanded" students of the 1960s now run the academic departments.

The Copenhagen interpretation was recently summarized as "Shut up and calculate!" That's blunt, but not completely unfair. It is, in fact, the right injunction for most physicists most of the time. The Copenhagen interpretation is the best way to deal with quantum mechanics for all practical purposes. It assures us that in our labs or at our desks we can use quantum mechanics without needless worrying about what's really going on. Copenhagen shows us that quantum mechanics is a fully consistent theory and sufficient as a guide to the physical phenomena around us. Good!

Perhaps, however, we wish more than an algorithm for computing probabilities. Classical physics provided more; it imparted a new worldview that changed our culture. It is, of course, a worldview we now know to be fundamentally flawed. Can it be that out there in our future there is a quantum impact on our worldview?

A Copenhagen Summary

☄ = Objector

☃ = Copenhagenist

☄ Quantum mechanics violates common *sense*. There must be something *wrong* with it!

☃ No. Never a wrong prediction. It works perfectly.

☄ The better it works, the sillier it looks! It's not logically consistent.

☃ Oh, you know that Einstein *tried* to show that. He gave up.

☄ But quantum mechanics says that little things have no properties of their own, that I actually create what I see by my *looking.*

☃ True. You perceive the basic idea quite clearly.

☄ But with only observer-created properties, little things have no physical *re-ality.* They're real only when they're being observed. That makes no sense!

☃ Don't worry about "reality," or about "making sense." Small, indirectly seen things are only models. Models need not make sense. Models only have to work. Large things are real enough. So everything's just fine.

☄ But a large thing is just a *collection* of little things, of atoms. To be consistent, quantum mechanics would have to say that *nothing* has a reality until it's observed.

☃ Oh, true, if you insist. But it doesn't matter.

☄ *Not matter?!* If quantum mechanics says my cat and my table aren't real until they're *looked* at, it's saying something *crazy.*

☃ No, it's all ok. You never actually *see* any craziness with big things. For all practical purposes, big things are *always* being looked at.

☄ For all *practical* purposes, sure. But what's the *meaning* of this observer-created reality?

☃ Science provides no meanings. Science just tells us what will happen. It just predicts what will be observed.

☄ I want more than a recipe for making predictions. If you say common sense is wrong, I want to know what's right.

🧍 But we've agreed that quantum mechanics is right. The Schrödinger equation tells everything that will happen, everything that can be observed.

🧍 I want to know what's really going on, I want the whole story!

🧍 The quantum mechanical description *is* the whole story. There's nothing else to tell.

🧍 Damn it! There's a real world out there. I want to know the truth about Nature.

🧍 Science can reveal no real world beyond what is observed. Anything else is just philosophy. *That's* the "truth"—if you must have one.

🧍 That's defeatist! I'll never be satisfied with such a superficial answer. You have science abandoning its basic philosophical goal, its mission to explain the physical world.

🧍 Too bad. But don't bother me with philosophy. I've got scientific *work* to do.

🧍 Quantum mechanics is manifestly absurd! I won't accept it as a final answer.

🧍 *(No longer listening.)*

11

Schrödinger's Controversial Cat

*The entire system would [contain] equal parts of living
and dead cat.*

—Erwin Schrödinger

When I hear about Schrödinger's cat, I reach for my gun.

—Stephen Hawking

By 1935, the basic form of quantum mechanics was clear. Schrödinger's equation was the new universal equation of motion. Although it was required only for objects on the atomic scale, quantum theory presumably governed the behavior of *everything*. The earlier physics, by then called "classical," was the much easier-to-use approximation adequate for macroscopic behavior.

Though quantum theory works perfectly, it says something weird: an object is created somewhere by our observation of it. We soon tell the story Schrödinger invented claiming that to be absurd. It's a story that resonates loudly today.

Most physicists nod at the quantum weirdness and then just accept the Copenhagen interpretation's permission to ignore it. If pressed, many will offer philosophical justifications for not being concerned. These usually involve the "nature of reality" just being different from what we can easily accept, because we have evolved in an essentially classical world.

In discussing any theory, some interpretation is implied. In all of what follows, when we refer to quantum theory, we mean its Copenhagen interpretation unless we say otherwise. As we've mentioned, the Copenhagen interpretation has been implicit in our discussion ever since we introduced the accepted meaning of the wavefunction.

Quantum theory has atoms and molecules not existing someplace until our observation creates them there. According to Heisenberg, they are not "real," just "potentialities." If unobserved atoms are somehow not physically real things, what does it say about things made of atoms? Chairs, for example? Is an unseen galaxy not really there? We're confronting the skeleton physics usually keeps in the closet.

Is it that quantum theory does not apply to big things? No: Quantum theory underlies all physics—we can't get to first base without quantum theory in dealing with such large-scale objects as lasers, silicon microchips, or the stars. Ultimately, the working of everything is quantum mechanical. But since we don't actually see the quantum strangeness with a big thing, Copenhagen insists, and most physicists pragmatically accept, that there is no reason to be bothered.

Schrödinger, however, *was* bothered: If quantum theory could deny the reality of atoms, it would also deny the reality of big things made of atoms. Schrödinger was sure something this crazy could not be Nature's universal law. We can imagine a conversation between a bothered Schrödinger and a pragmatic young colleague.

SCHRÖDINGER: The Copenhagen interpretation is a cop-out. Nature is trying to tell us something. Copenhagen is telling us not to listen. Quantum theory gives an absurd worldview.

COLLEAGUE: But sir, your theory works perfectly. No prediction has ever been wrong. So everything's okay!

S: Come now, I look and I find an atom someplace. The theory says that just before I looked it wasn't there—it didn't exist at that place. It didn't exist *anyplace!*

C: That's right. Before you looked to see where it was, it was a wavefunction, just probability. The atom *didn't* exist at any particular place.

S: You're saying my looking *created* the atom at the place I found it?

C: Well, yes, sir. That's what your theory says.

S: That's silly solipsism. You're denying the existence of a physically real world. This chair I'm sitting on is a very real chair.

C: Oh, of course, Professor Schrödinger, your chair's real. Only the properties of small things are created by observation.

S: You're saying quantum theory applies only to small things?

C: No, sir, in principle your equation works for everything. But it's impossible to do an interference experiment with a big thing. So for all practical purposes there's no reason to worry about the reality of big things.

S: A big thing is just a collection of atoms. If an atom doesn't have physical reality, a collection of them can't be real. If quantum theory says that the real world is created by our looking at it, the theory's absurd!

By a logical technique called *reductio ad absurdum,* or reduction to an absurdity, Schrödinger told a story to argue that quantum theory, or at least its Copenhagen interpretation, led to an absurd conclusion. Decide for yourself whether to accept his argument. But wait until we also present the standard counter to his reasoning.

The Cat in the Box Story

Forgive our once again repeating the box-pair story. It's the first step in Schrödinger's argument. Recall our atom hitting a partially reflecting/partially transmitting mirror and ending up with half its waviness captured equally in each of two separate boxes. According to quantum theory, the atom does not exist in one particular box before you find a whole atom to be in one of the boxes. The atom is in a superposition state simultaneously in both boxes. Upon your looking into one box, the superposition state waviness collapses into one single box. You will randomly find either a *whole* atom in that one box *or* that box will be empty. (You can't choose which!) If you find the one box empty, the atom will be found in the other box. But with a set of box-pairs, you could have produced an "interference pattern" demonstrating that *before* you looked, the atom had been simultaneously in each box.

Our version of Schrödinger's story takes off from here. Suppose now that, before we send in an atom, one of the boxes of the pair is not empty. It contains a Geiger counter designed to "fire" if an atom enters its box. In firing, this Geiger counter moves a lever to pull the cork from a bottle of hydrogen cyanide. There's also a cat in the box. The cat will die if the poisonous cyanide escapes its bottle. The entire content of the boxes, the atom, the Geiger counter, the cyanide, and the cat, is isolated and unobserved.

We immediately note that Schrödinger never contemplated actually endangering a cat. This was a thought experiment. He referred to the apparatus as a "hellish contraption."

Now, Schrödinger argued, a Geiger counter is just a bunch of ordinary atoms, albeit a complex and well-organized collection. It is governed by the same laws of physics that govern the atoms it's made of—by quantum mechanics. Presumably the same is true for the cat.

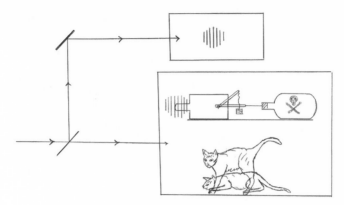

Figure 11.1 Schrödinger's cat

Since the waviness of the atom split equally at the semitransparent mirror, half went into the box with the Geiger counter and the cat and half into the other box. As long as the system is isolated, the atom is in a superposition state we can describe as in the box with the Geiger counter and, simultaneously, in the empty box. To be succinct we say that the atom is simultaneously in both boxes.

The unobserved Geiger counter must therefore also be in a superposition state. It is both fired and, simultaneously, unfired. The cork on the cyanide bottle must be both pulled and not pulled. The cat must be both dead and alive. This is, of course, hard to imagine. Impossible to imagine, perhaps. But it's just a logical extension of what quantum theory is telling us.

We show quantum theory's version of our yet-unobserved cat and the rest of Schrödinger's "hellish contraption" with a mixed-metaphor image. We represent the atom by its wavefunction in both boxes. Since the wavefunctions of Geiger counters and cats are too complicated to display, we just picture the Geiger counter both fired and unfired (lever both up and down), the cyanide cork both pulled and not pulled, and the cat simultaneously dead and alive.

What if you now look into the box to see whether the cat is dead or alive? Back when only an atom was in a superposition state in our box pair, any look into a box collapsed the atom totally into one box or the other. Here, a look collapses the wavefunction of the entire system.

The theory must predict a self-consistent situation. If you find the cat dead, then the Geiger counter will have fired, the cork on the cyanide bottle will be pulled, and the atom will be in the box with the cat. If you find the cat alive, the Geiger counter will not have fired, the cyanide bottle will be corked, and the atom will be in the other box.

But, according to quantum theory, before you looked, the atom was not in one box or the other. It was in a superposition state simultaneously in both boxes.

Therefore, assuming cats are not entities beyond the laws of physics, before you looked, the cat was in a superposition state equally alive and dead. It was not a sick cat. It was a perfectly healthy cat and a stone-dead cat at the same time.

Though the alive or dead condition of the cat did not exist as a physical reality until observed, the existence of the cat in the box was a reality. But only because that existence was observed by whoever put the cat there.

Since your looking collapsed the superposition state of the cat, are you guilty of killing the cat if you find it dead? Not really, assuming you didn't arrange the "hellish contraption" in the first place. You could not have chosen how the wavefunction of this entire system would collapse. The collapse into either the living or the dead state was random.

Here's something to ponder: Suppose the cat was placed in the box and the atom sent into the mirror system eight hours before you looked. The system evolves unobserved during those eight hours. If you find the cat alive, since it has gone eight hours without eating, you find a hungry cat. If you find a dead cat, an examination by a veterinary forensic pathologist would determine the cat to have died eight hours ago. Your observation not only creates a current reality, it also creates the history appropriate to that reality.

You might consider all this absurd. Precisely Schrödinger's point! He concocted his cat story to argue that, taken to its logical conclusion, quantum theory, at least its Copenhagen interpretation, was absurd. Therefore, he claimed, it must not be accepted as a description of what's really going on.

The idea of a cat simultaneously alive and dead was, of course, as ridiculous to other physicists as it was to Schrödinger. But few worried about Schrödinger's demonstration of the theory's absurdity. The theory worked too well for mere absurdity to be a serious challenge.

We return in a moment to the controversy Schrödinger's story still raises. But first, if the cat is simultaneously alive and dead, can we somehow *see* it that way? No. Although we sketched a superposed live and dead cat (figure 11.1), you'll never see a cat like that. Observation collapses the whole system putting the cat into either the living or the dead state. But what about just a peek? Can a tiny peek collapse the wavefunction of a whole cat?

Consider the tiniest possible peek. That could be bouncing a single photon off the cat through tiny holes in the box. With a single photon you can't learn much. But if that photon were blocked, telling us that the cat was standing, and therefore alive, it would collapse the superposition state into the living state. Quantum theory tells us that *any* look, anything in fact that provides information, collapses the previously existing state. There's no immaculate perception.

Wait a minute! Can't the *cat* observe whether or not the cyanide cork has been pulled, and therefore whether the atom entered its box? Don't cats qualify as observers and collapse wavefunctions? Well, if cats, what about mosquitoes? Viruses? Geiger counters? How far down do we go? The two smart cats, the ones that live with each of us, are certainly conscious observers. But how can we be sure?

Strictly speaking, all you know for sure is that you are a wavefunction-collapsing observer. The rest of us may merely be in a superposition state governed by quantum mechanics and are collapsed to a specific reality only by your observation of us. Of course, since the rest of us look and act more or less like you, you trust that we also qualify as observers. (We soon discuss the "many-worlds" interpretation of quantum mechanics, which suggests that we are all in superposition states.)

Although it's just a logical extension of what quantum theory says, solopsistic talk like this seems just plain silly. Nevertheless, some physicists seriously consider the possibility that quantum mechanics hints of a mysterious connection of *conscious* observation with the physical world. Eugene Wigner, one of the later developers of quantum theory and a winner of a Nobel Prize in physics, created a version of the cat story suggesting an even stronger involvement of the conscious observer with the physical world than Schrödinger's story.

Instead of a cat, Wigner contemplated having a friend stay unobserved in one of the boxes, a room. No cyanide this time. The Geiger counter firing just goes "click." His friend would mark an "X" on a pad if she hears a click. Since Wigner could not imagine that he collapsed her superposition state wavefunction when he opened the door and looked at her pad, he assumed that any human has status as an observer. Wigner speculated that collapse happens at the very last stage of the observation process, that his friend's human consciousness collapsed the physical system's wavefunction. Going even further, he speculated that human conscious awareness might actually "reach out"—in some unexplained way—and change the physical state of a system.

You can't prove otherwise. All we know is that someplace on the scale between big molecules and humans there is this mysterious process of observation and collapse. Conceivably, it's indeed at the last step, at awareness. We explore some seriously proposed ideas regarding this in later chapters.

The Response to Schrödinger's Story

We've entered emotional territory. Most physicists squirm when their discipline is associated with "soft" subjects such as consciousness. Some are even infuriated when Schrödinger's cat story is told. Stephen Hawking claims to "reach for my gun."

We'll give a more or less standard response to Schrödinger's story. First, though, a "truth in advertising" statement: Our sympathies are with Schrödinger's concern. Were that not so, we'd not be writing this book. Nevertheless, we'll present as strong an argument as we can that Schrödinger's cat story and the discussion of conscious observation are irrelevant and misleading. For the next several paragraphs we take that point of view.

Schrödinger's argument fails because it rests on the assumption that macroscopic objects can remain unobserved in a superposition state. For all practical purposes, any macroscopic object is constantly "observed." It can't be isolated; it's always in contact with, entangled with, the rest of the world. And that entanglement is observation!

It is ridiculous to imagine that a cat could be isolated. Every macroscopic object anywhere near the cat observes the cat. The photons emitted by the warm walls of the box, for example. Take an extreme example: the moon! The moon's gravity, which pulls on the oceans to raise the tides, also pulls on the cat. That pull would be slightly different for a standing, alive cat than for a lying, dead cat. Since the cat pulls back on the moon, the path of the moon is slightly altered depending on the position of the cat. It is easy to calculate that in a tiny fraction of a millionth of a second the cat's wavefunction would be completely entangled with the moon's, and thus with the tides and thus with the rest of the world. This entanglement *is* an observation. It collapses the superposition state of the cat in essentially no time at all.

Even looking back at the earliest stage of Schrödinger's story, you can see how absolutely meaningless it is. When an atom is sent into Schrödinger's boxes, its wavefunction becomes entangled with the enormously complex wavefunction of the macroscopic Geiger counter. The atom is therefore "observed" by the Geiger counter. Since something as big as a Geiger counter can't, for all practical purposes, be isolated from the rest of the world, the rest of the world observes the atom. Entanglement with the world *constitutes* observation, and the atom collapses into one box or the other as soon as its wavefunction enters the box pair and encounters the Geiger counter. And the cat is either dead or alive. Period!

Even if you (needlessly!) bring consciousness into the argument, big things are constantly being observed if only because they are always in contact with something that is observed by somebody conscious.

If such arguments don't convince you that there's nothing to the cat story, here is a final put-down of Schrödinger's claim to having demonstrated a problem with quantum theory. Do the experiment! You'll always get the result quantum theory predicts; you'll always see either an alive or a dead cat. The Copenhagen interpretation makes it clear that the role of science is to predict the results of observations, not to discuss some "ultimate reality." Predictions of what will happen are all we ever need. You'll find the cat alive half the time and dead half the time. Conscious observation is irrelevant. The cat story raises a misleading nonissue.

We now no longer speak as a responder to Schrödinger's argument and return to our own voice. Schrödinger was, of course, fully aware of the difficulty of isolating anything as large as a cat. He would argue that such practical problems are beside the point. Since quantum theory admits no boundary between the small and the large, in principle any object can be in a superposition state. He (along with Einstein) rejected as defeatist the Copenhagen claim that the role of science is merely to predict the results of observations, rather than to explore what's really going on.

No matter which side of this argument you favor, there are physicists who'd agree with you.

Schrödinger's Cat Today

Seven decades after Schrödinger told his story, conferences almost every year address the quantum enigma and often include discussions of consciousness. Reference to the cat story in professional physics journals increases. Two examples: An article, " 'Schrödinger Cat' Superposition State of an Atom," demonstrated such a state. In another, "Atomic Mouse Probes the Lifetime of a Quantum Cat," the "mouse" is an atom and the "cat" is an electromagnetic field in a macroscopic resonant cavity. Though these are serious and expensive physics projects, the titles illustrate how our discipline is inclined to approach the weirdness of quantum mechanics with a bit of humor.

Speaking of humor, here is a cartoon from the May 2000 issue of *Physics Today*, the most widely distributed journal of the American Institute of Physics. It would not likely have been published twenty years ago.

Though the mysterious aspects of quantum mechanics are still hardly discussed in physics courses, the issue increasingly intrudes. A recent quantum mechanics text has a picture of a live cat on the front cover and a dead cat on the back—though

Figure 11.2 Drawing by Aaron Drake, 2000.
© American Institute of Physics

there is very little talk of the cat inside. (We suspect the publishers, not the author, chose the cover design.)

Experimental studies of the mysterious aspects of quantum mechanics that would not have been proposed years ago, and would not have been funded if proposed, now get considerable attention. Increasingly large objects are being put into superposition states, put into two places at the same time. Austrian physicist Anton Zeilinger has done this with large molecules containing seventy carbon atoms—football-shaped "buckyballs." He's now setting up to do the same thing with mid-sized proteins. At a recent conference he was asked: "What's the limit?" His answer: "Only budget."

Truly macroscopic superpositions containing many billions of electrons have been demonstrated where each electron is simultaneously moving in two directions. Bose-Einstein condensates have been created in which each of several thousand atoms is spread over several millimeters. A recent American Institute of Physics news bulletin bore the headline "3600 Atoms in Two Places at Once." It gets harder to dismiss Schrödinger's concern by saying the weirdness is only evident with the small things we never actually see.

Perhaps hardest to accept is the claim that your observation not only creates a present reality but also creates a past appropriate to that reality—that, when you looking collapsed the cat to being either alive or dead, you also created the history appropriate to an eight-hour-hungry cat or to an eight-hour-dead cat.

The "delayed-choice experiment" suggested by Princeton quantum cosmologist John Wheeler comes closest to testing this idea. Consider our original box pairs without the cat or the rest of Schrödinger's "hellish contraption." If we choose an interference experiment, the atom, on its encounter with the semitransparent mirror, would have had to "decide" to come to the boxes on both paths. If we choose a look-in-the-box experiment, the atom would have had to come on a single path. It would have had to make a different "decision" back at the semitransparent mirror. Quantum theory is saying that our later choice of observation creates the atom's earlier history—we cause something *backward in time.*

Backward in time is hard to accept. Therefore, maybe what really happened is that our earlier mechanical setting up of the equipment, depending on which observation we were *going* to make, somehow affected the atom's later decision at the semitransparent mirror. Even if we can't understand how such mechanical setting up could possibly affect the atom's later behavior, at least that would avoid the future appearing to cause the past. To test such a possibility, Wheeler suggested an experiment where the choice of experiment was delayed until after the object had already passed the semitransparent mirror, after its "decision" there had already been made.

If observation did not, in fact, create history, such a "delayed choice" might well produce a result different than that predicted by quantum theory. In 1987 Wheeler's experiment was done, though with photons instead of atoms. Unfortunately for the experimenters, quantum theory's prediction, that the later choice of experiment determined what the photon did earlier at the semitransparent mirror, was confirmed. There would have been a quick Nobel Prize had they produced the first-ever disconfirmation of a quantum theory prediction.

Too bad Schrödinger isn't around to see the increasing interest in his cat. He felt that Nature was trying to tell us something and that physicists should look beyond a pragmatic acceptance of quantum theory. He'd agree with John Wheeler: "Somewhere something incredible is waiting to happen."

12

Seeking a Real World
EPR

I think that a particle must have a separate reality independent of the measurements. That is, an electron has spin, location and so forth even when it is not being measured. I like to think the moon is there even if I am not looking at it.

—Albert Einstein

Schrödinger told his cat story to show that quantum theory denied the existence of a physically real world, that quantum theory claimed that observation created the observed reality. That seems crazy. Indeed, if someone on trial convinced the jury that he believed that his looking created the physical world, the jury would likely accept a plea of insanity.

The Copenhagen interpretation is, of course, more subtle. It claims only that objects of the *microscopic* realm lack reality before they are observed. Moons, chairs, and cats are real—for all practical purposes. And that, according to Copenhagen, should be good enough. But that was not good enough for Einstein, who wanted to know "God's thoughts."

At the 1927 Solvay conference, Einstein, by then the world's most respected scientist, turned thumbs down on the newly minted Copenhagen interpretation. He insisted that even little things have reality, whether or not anyone is looking. And if quantum theory said otherwise, it had to be wrong. Niels Bohr, the Copenhagen interpretation's principal architect, rose to its defense. For the rest of their lives Bohr and Einstein debated as friendly adversaries.

Evading Heisenberg

Quantum theory has an atom being either a spread-out wave or a concentrated particle. If, on the one hand, you look and see it come out of a single box (or through a single slit), you show it to be a compact particle. On the other hand, it can participate in an interference pattern that shows it to be an extended wave—an apparent contradiction. But the theory is protected from refutation by the Heisenberg uncertainty principle, which shows that looking to see through which slit an atom comes kicks it hard enough to blur any interference pattern. So you thus can't demonstrate a contradiction.

To argue that quantum theory led to an inconsistency and was therefore wrong, Einstein attempted to show that even though an atom participated in an interference pattern, it *actually* came through a single slit. To demonstrate this he had to evade the uncertainty principle. (Ironically, Heisenberg attributed his original idea for the uncertainty principle to a conversation with Einstein.) Here's Einstein's challenge to Bohr at the 1927 Solvay conference:

Send atoms toward a two-slit barrier one at a time. Let the barrier be movable, say, on a light spring. Consider the simplest case, an atom that landed in the central maximum of the interference pattern (point A in figure 12.1). If that atom happened to come through the bottom slit, it had to be deflected upward by the barrier. In reaction, the atom would kick the barrier downward. And vice versa if the atom went through the top slit.

By measuring the movement of the barrier after each atom had passed, one could know through which slit it went. This measurement could be made even

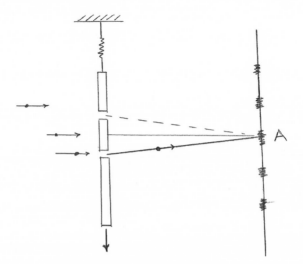

Figure 12.1 Atoms fired one at a time through a
movable two-slit barrier

after the atom was recorded as part of an interference pattern on a photographic film. Since one could thus know through which slit each atom came, quantum theory was wrong in explaining the interference pattern by claiming each atom to be a wave passing through *both* slits.

Bohr readily pointed out the flaw in Einstein's reasoning: For Einstein's demonstration, one would have to know simultaneously both the barrier's initial position and any motion it might have had. The uncertainty principle limits the accuracy with which both position and motion can be simultaneously known. With simple algebra, Bohr was able to show that this uncertainty would be large enough to foil Einstein's demonstration.

Three years later at another conference, Einstein proposed an ingenious thought experiment claiming to violate a version of the uncertainty principle by determining both the time a photon exited a box and its energy, both with arbitrarily great accuracy. This one stumped Bohr through a sleepless night. But in the morning he embarrassed Einstein by showing that in his attempt to evade the uncertainty principle he violated his own general theory of relativity. Years later, Bohr revisited this triumph with a nuts-and-bolts caricature of Einstein's photon-in-a-box experiment illustrating his general rule preventing such refutation: In any quantum experiment one must consider the macroscopic apparatus actually used.

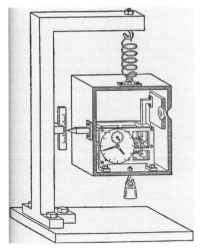

Figure 12.2 Bohr's drawing of Einstein's clock-in-the-box thought experiment. Courtesy HarperCollins

Bohr's refutation of Einstein's thought experiments has been questioned. In chapter 10 we quoted Bohr saying, "measuring instruments [must be] rigid bodies sufficiently heavy to allow a completely classical account of their relative positions and velocities." Was Bohr's application of the quantum mechanical uncertainty principle to the *macroscopic* slit barrier and the photon-box apparatus consistent with his requiring a "completely classical account" of the macroscopic measuring instruments? Bohr at least seems to agree that quantum theory, and thus the question of observer-created reality, applies in principle to the big as well as the small. Only for all practical purposes do large things behave classically. In any event, Bohr's arguments convinced Einstein that the theory was at least consistent and that its predictions would always be correct. A humbled Einstein went home from the conference to concentrate on general relativity, his theory of gravity, or so Bohr assumed.

A Bolt from the Blue

Bohr was wrong to think Einstein had abandoned his attempt to fault quantum theory. Four years later (in 1935), a paper by Einstein and two young colleagues, Boris Podolsky and Nathan Rosen, arrived in Copenhagen. An associate of Bohr tells that "this onslaught came down upon us like a bolt from the blue. Its effect on Bohr was remarkable . . . as soon as Bohr heard my report of Einstein's argument, everything else was abandoned."

The paper, now famous as "EPR" for "Einstein, Podolsky, and Rosen," did not claim that quantum theory was wrong, just that it was incomplete. Quantum theory supposedly denied a physically real world, and thus required an observer-created reality, only because it was not the whole story.

EPR would show that you could, in fact, know a property of an object *without* observing it. That property, they argued, was therefore not observer created. The property was a physical reality that the "incomplete" quantum theory did not include. Here's a classical analogy—one that stimulated Einstein's EPR argument:

> Consider two identical railroad cars latched together but pushed apart by a strong spring. Suddenly unlatched, they take off at the same speed in opposite directions. Alice, on the left (figure 12.3), is closer to the cars' starting point than is Bob, on the right. Observing the position of the car passing her, Alice immediately knows the position of Bob's car. Having no effect on Bob's car, Alice did not create its position. Not yet having observed his car, Bob did not create its position. Since the position of Bob's car is not observer created, it was always physical reality.

The conclusion arrived at in this Alice and Bob story is so obvious that it seems trivial. But replace the railroad cars with two atoms flying apart, and quantum theory tells us that their positions are created by observation.

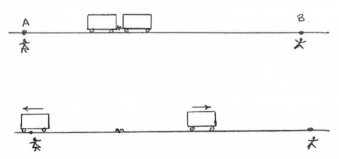

Figure 12.3 A classical analogy of the EPR argument

Polarized Light

Unfortunately, there is a problem converting the easily visualized railroad car analogy to the quantum mechanical situation: The uncertainty principle forbids knowing both the initial speed and position of the cars well enough. We skip EPR's ingenious but hard-to-visualize mathematical trick and go to David Bohm's polarized photon version of EPR. It is worth exploring polarized photons because the mysterious quantum influences revealed by EPR-type experiments are mostly demonstrated with photons. Those "spooky" influences are the subject of our next chapter.

In the next few pages, we go over some physics of polarized light and polarized photons so that we can then present the profound EPR argument compactly. Even if you just skim these details of polarized photons or simply skip down to the section headed "EPR," you can still appreciate Einstein's argument.

Light, recall, is a wave of electric (and magnetic) field. Light's electric field can point in any direction perpendicular to the light's travel. In our sketch (figure 12.4), the light is going away from the reader, with its electric field in the vertical direction. Such light is "vertically polarized." The other sketch shows a horizontally polarized light wave. The direction of light's electric field is its direction of polarization.

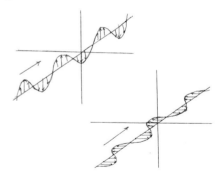

There is, of course, nothing special about the vertical and horizontal directions—other than that they are perpendicular to each other. It's just conventional to speak of "vertical" and "horizontal."

Figure 12.4 Vertically and horizontally polarized light

The polarization of light from the sun or a light bulb—most light, in fact—varies randomly. Such light is "unpolarized." Certain materials allow the passage only of light polarized along a particular molecular alignment in the material. Such "polarizers" in sunglasses cut down glare by not transmitting the largely horizontally polarized light reflected from horizontal surfaces such as roads or water. But we will describe a different sort of polarizer.

The polarizer actually used in the experiments we describe here is a transparent crystal of the mineral calcite. It doesn't absorb light; it just sends light of different polarizations on different paths. Light polarized parallel to the crystal's "axis" is sent on Path 1, and light polarized perpendicular to that axis is sent on Path 2.

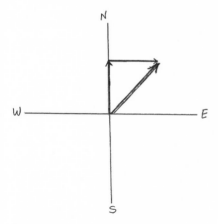

Figure 12.5 Traveling northeast as the sum of traveling north and then traveling east

Light polarized at an angle other than parallel or perpendicular to the crystal's axis can be thought of as a sum of parallel and perpendicular polarized components. (It's the way a trip northeast can be thought of as the sum of a trip with one component north and another east.) The parallel component of the light goes on Path 1, and the perpendicular on Path 2. The closer the polarization is to parallel, the more light goes on Path 1.

Polarized Photons

Light is a stream of photons. Photon detectors can count individual photons. They can count millions per second. Our eye, incidentally, can detect light as dim as ten photons per second.

Light polarized parallel to the crystal axis is a stream of parallel-polarized photons. Each of them goes on Path 1 to be recorded by the photon detector on Path 1. Similarly, the Path 2 detector will record every photon polarized perpendicular to the crystal axis. The photons of ordinary unpolarized light are randomly polarized. On encountering the calcite crystal, each is recorded by either the Path 1 or the Path 2 detector. In our sketch (figure 12.6) we show a photon as a dot, its polarization as a double-headed arrow, the calcite crystal as a box, and the detectors as D1 and D2.

We must say a bit more about photons polarized at an angle other than parallel or perpendicular to the axis of our calcite crystal. Such photons have a certain probability of being recorded by the Path 1 or the Path 2 detector. A photon polarized at forty-five degrees to the crystal axis, for example, has equal probability of being recorded by either detector. The closer the polarization is to being parallel to the crystal axis, the greater its probability of being recorded by the Path 1 detector.

Note that we are careful *not* to say that a photon at some angle other than parallel or perpendicular actually went on either path. It goes into a superposition state traveling simultaneously on both paths. A photon polarized at forty-five degrees, for example, goes equally on both paths. *But we never see partial photons.* A detector clicks and records a whole photon, or it remains silent, indicating that no photon came.

The situation for the forty-five degree photon is analogous to our atom in a box pair. Looking at either path with a photon detector, we find a whole photon

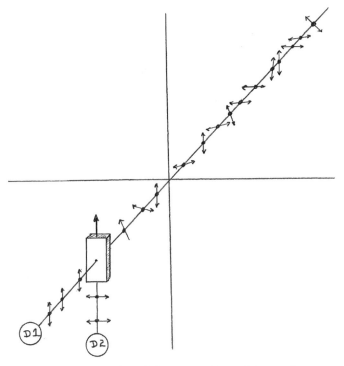

Figure 12.6 Polarized photons sorted by a calcite crystal

or no photon. Only when one of the photon detectors clicks and records a photon does the photon's superposition state collapse.

(You could demonstrate that a photon was in a superposition state on both paths by something analogous to an interference experiment. Instead of having a detector on each path, have mirrors on each path reflect photons through a second calcite crystal that recombines the parallel and perpendicular components of the photon to reproduce the original forty-five degree photon. Change the length of *either* path, and you change the polarization of the resulting photon. That demonstrates that it came on both paths.)

In saying photon detectors record photons, we're taking a Copenhagen interpretation stance. We're regarding the macroscopic photon detectors as observers. When one of the detectors records the presence of a photon on a particular path, the superposition state collapses, and that photon is absorbed. What remains is the detector's record of the photon.

Einstein, of course, accepted none of this superposition-state business. For him, photons and atoms were as real as railroad cars. For him, a photon encountering a calcite crystal actually went either on Path 1 or Path 2—not on both.

Before we come to the EPR argument making this point, we must tell of photons in the "twin state."

Twin-State Photons

Atoms can be raised to excited states from which they return to the ground state by two quantum jumps in rapid succession. In such a cascade, the atom releases two photons. Since there's nothing special about a particular direction in space, the polarizations observed for the photons will be completely random.

But here's the crucial point: For certain atomic states, the two photons that fly off in opposite directions will always display the same polarization. If, for

example, the photon going off to the left is observed to have vertical polarization, its twin, the photon going off to the right, will then also be vertical.

Figure 12.7
A two-photon cascade

We must, of course, ensure that the two photons came from the same atom. That's not too hard with fast electronic photon counters. If two photons arrive at equidistant polar-

izers at precisely the same time, they must have been emitted by the same atom and in fact be twins.

The reason twin-state photons always exhibit the same polarization doesn't matter here. (It's required to conserve angular momentum, and in this case the initial and final atomic states have the same angular momentum.) The important thing is that it is demonstrably true that their polarizations are always observed as identical.

Back to Alice and Bob, with photons instead of railroad cars: A twin-state photon source is between Alice, on the left, and Bob, on the right (figure 12.8). They each observe the polarization of twin-state photons with the axis of their polarizers oriented at the same angle. Their Path 1 and Path 2 photon detectors randomly click and record the arrival of a photon polarized parallel or perpendicular to their polarizer crystal axis. However, whenever Alice observes her Path 1 detector to record a photon, Bob *always* finds its twin to go on his Path 1. Whenever Alice observes her Path 2 detector to record a photon, Bob finds its twin to go on his Path 2.

Since the photons are twins, it might not seem strange that they always exhibit the same polarization. Let's play with an analogy: It is not surprising that identical-twin boys exhibit the same eye color. Identical twins are created with the same eye color. Consider another property of the twins: the color of sock they choose to wear each day. Suppose whenever one twin chooses green, the other chooses green that day, even though neither twin had information about

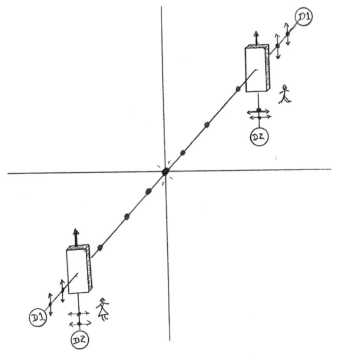

Figure 12.8 Alice and Bob with twin-state photons

his brother's sock-color choice. That would be strange because the twins were not created with the same daily sock-color choice.

Can we explain twin-state photons exhibiting identical polarization because they were created with, say, vertical polarization, the way the twins were created with blue eyes? Not according to quantum theory, which claims no properties exist before they are observed. In particular, the polarization of a photon does not exist as a physical reality before its observation. That's why we put no polarization arrows on the photon twins on their way to Alice and Bob.

(Incidentally, in the case of ordinary unpolarized light in figure 12.6, we did put arrows on the incident photons because we could assume that the atoms emitting them were observed—by the macroscopic filament of the light bulb that emitted them, for example.)

So here's the important point: If Alice's photon happens to be recorded by her Path 1 detector, its twin will surely exhibit the same polarization by being recorded by Bob's Path 1 detector—even though Bob's photon could not possibly have received any information of the behavior of its twin at Alice's polarizer. This is like the strange sock-color choice of the twin boys.

Nothing Alice did could physically affect the photon detected at Bob's polarizer. Alice and Bob could even be in different galaxies and observe the twin photons at essentially the same time. How, then, did the two photons acquire identical polarization on the observation of the polarization of one of them?

Quantum theory's explanation for this behavior is hard to believe, it's worth repeating: Twin-state photons do not have a particular polarization until the polarization of one of them is observed. Twin-state photons are entangled in a state of identical polarization but have no particular polarization. It is the observation of the polarization of one of the photons as being, say, vertical that instantaneously collapses *both* photons to vertical polarization.

It's not the fact that twin-state photons exhibit identical polarization that is weird. The weird thing is quantum theory's explanation of that fact: that there is no physical reality to any property until it is observed.

Quantum theory's denial of physical reality bothered Einstein far more than its randomness. His remark, "God does not play dice," is often quoted. But the less-easily understood quotation we headed this chapter with, "I like to think the moon is there even if I am not looking at it," captures his more serious concern.

EPR

The EPR paper that arrived in Copenhagen as a "bolt from the blue" was titled: "Can Quantum-Mechanical Description of Physical Reality Be Considered Complete?" (Historians have attributed the missing "the" to the paper being worded by Podolsky, whose native Polish does not include articles.) The EPR paper talked of a complex combination of position and momentum of particles instead of photon polarization. But we discuss it in the simpler, and more modern, way by talking in terms of photons.

Quantum theory's description of the twin-state photons themselves does not include their actual polarization as a physically real property. Yet quantum theory claimed to be a complete theory of the phenomena.

To dispute this claim of completeness, EPR had to say what constituted a "physical reality." Defining reality has been a philosophical issue at least since Plato's day. But EPR did not need to define reality in general, they merely needed a sufficient condition for something to be a physical reality. If that physical reality were not described by the theory, the theory would be incomplete. Here's the condition offered by EPR:

If without in any way disturbing a system, we can predict with certainty
. . . the value of a physical quantity, then there exists an element of physi-
cal reality corresponding to this physical quantity.

Let's say the same thing in other words: If a physical property of an object
can be known *without* its being observed, then that property could not have been
created by observation. If it wasn't created by its observation, it must have existed
as a physically reality before its observation. EPR needed to display only one such
property to show quantum theory to be incomplete.

Quantum theory has twin-state photons in a state of identical polarization
but with no *particular* polarization. Observation of the polarization of one photon
supposedly creates the physically real polarization of both photons.

Back to Alice and Bob with their polarizers and photon detectors: This time
Alice is a bit closer to the twin-state photon source than is Bob. She therefore
receives her photon before Bob receives its twin. Suppose just now she observes
a photon to be polarized parallel to their agreed-upon axis direction; it was recorded by her Path 1 detector. She immediately knows that its twin on its way
to Bob is parallel polarized. She knows it will go on Path 1 when it reaches Bob's
polarizer.

In fact, it would be possible for Bob to trap his photon in a pair of boxes, one
of which is fed by Path 1 and the other by Path 2. After his photon is trapped,
Alice could telephone him and tell him with certainty in which box he would
find his photon.

Alice could not have physically disturbed Bob's photon. It started at the
photon source and moved away from her at the speed of light. Since nothing can
travel faster than that, nothing Alice could send at Bob's photon could ever catch
up with it. When Alice observed her photon, Bob's hadn't yet gotten to him; he
therefore could not have disturbed it.

Neither Alice nor Bob, nor anybody, observed the polarization of Bob's photon. Yet its unobserved polarization can be known with certainty.

That does it! Alice's knowing with certainty the polarization of Bob's photon—without in any way disturbing it—meets EPR's criterion for the polarization of Bob's photon being a physical reality. Since quantum theory does not
describe this physical reality, EPR claimed the theory to be incomplete. The EPR
paper concluded with the authors stating their belief that a complete theory is
possible. Such a complete theory would presumably give a reasonable picture of
the world, a world existing independently of its observation.

Bohr's Response to EPR

When he received the EPR paper, almost a decade after the Copenhagen interpretation was developed, Bohr had not yet fully realized the implications of quantum theory, in particular, the implication to which EPR objected: that observation, in and of itself, *without any physical disturbance,* can instantaneously affect a remote physical system.

Bohr recognized Einstein's "bolt from the blue" as a serious challenge to quantum theory, and he worked furiously for weeks to develop a response. A few months later he published a paper with exactly the same title as EPR: "Can Quantum Mechanical Description of Physical Reality Be Considered Complete?" (He even left out the "the.") While EPR's answer to the paper's title question was "no," Bohr's was a firm "yes."

Here's an extract from Bohr's long response to EPR. It carries the essence of his complex argument:

> [The] criterion of physical reality proposed by Einstein, Podolsky and Rosen contains an ambiguity as regards to the meaning of the expression "without in any way disturbing a system." Of course there is in a case like that just considered no question of a mechanical disturbance of the system under investigation during the last critical stage of the measurement procedure. But even at this stage there is essentially the question of an *influence on the very conditions which define the possible types of predictions regarding the future behavior of the system.* Since these conditions constitute an inherent element of the description of any phenomenon to which the term "physical reality" can be properly attached, we see that the argument of the mentioned authors does not justify their conclusion that the quantum-mechanical description is essentially incomplete.

Let's analyze this refutation of EPR. First of all, Bohr did not fault the logic of the EPR argument. He rejected their starting point, their condition for something being a physical reality.

EPR's reality condition tacitly assumes separability: If two objects exert no physical force on each other, what happens to one cannot in any way "disturb" the other. Let's be specific regarding Alice and Bob's twin-state photons: Alice, by observing her photon, cannot exert a physical force on Bob's photon, which is moving away from her at the speed of light. Therefore, according to EPR, she cannot have any effect on it.

Bohr agreed that there could be no "mechanical" disturbance of Bob's photon by Alice's observation. (*All* physical forces are included in Bohr's term "mechanical.") He nevertheless maintains that even without a physical disturbance, Alice's remote observation instantaneously "influences" Bob's photon. And, according to Bohr, this constitutes a disturbance violating the EPR condition for reality. Only after Alice observed her photon as, say, polarized parallel was Bob's photon polarized parallel.

Did Alice's observation physically affect Bob's photon? Can what is done at a distant place, even on a faraway galaxy, instantaneously cause something to happen here? Certainly no physical force affected Bob's photon. What, then, did Alice's observation do to Bob's photon? Strictly speaking, we should not say her observation "affected" Bob's photon or "caused" its behavior because no physical force was involved. We properly use the mysterious term sanctified by Bohr: Alice "influenced" its behavior.

Notice, incidentally, that Alice cannot communicate any information to Bob by her observations; he always sees a series of random photon polarizations. Only when Alice and Bob come together and compare their results do they see the remarkable correlation: Whenever she saw a photon parallel to their agreed-upon detector orientation, so did he; whenever she saw a perpendicular photon, so did he.

To defend quantum theory in spite of its "nonphysical" aspect, Bohr redefined the goal of science. That goal is not, he later claimed, to describe Nature, but only to describe what we can say about Nature. In his earlier debates with Einstein, Bohr argued that any observation *physically* disturbs what you observe by an amount enough to prevent any experimental refutation of quantum theory. This has been called a "doctrine of physical disturbance." Since Alice's observation changes what can be correctly *said* about Bob's photon, Bohr's response to EPR has been called a "doctrine of semantic disturbance."

Is all this confusing? You bet! There is no way that EPR and Bohr's response to it can be correctly stated that does not either confuse or sound mystical.

Einstein rejected Bohr's response. He insisted that there was a real world out there, and science must explain it. A photon displayed a particular polarization not because some other object was observed but because that photon actually had a physical property determining its polarization. If that property, later called a "hidden variable," was not in quantum theory, the theory was incomplete. He derided Bohr's "influences" as being "voodoo forces" and "spooky interactions." He could not accept such things as part of the way the world works, saying: "The Lord God is subtle, but malicious He is not."

We should be clear that Bohr and Einstein would agree on the actual results of an EPR experiment, the Alice and Bob observations we described. They would just interpret those results differently.

It is fair to speculate why they so strongly held to their philosophical positions. Einstein was forever dubious about quantum theory; Bohr was its staunchest defender. Recall that for almost twenty years the physics community rejected young Einstein's quantum proposal that light came as photons—it was called "reckless." In contrast, Bohr's early work on quantum mechanics brought him immediate acclaim. Did their early professional experiences shape their lifelong attitudes?

Einstein thought physicists would reject Bohr's arguments refuting EPR. He was wrong. Quantum theory worked too well. It provided a basis for rapid advance in physics and its practical applications. Working physicists had little inclination to deal with philosophical issues.

In the two decades he lived after EPR, Einstein never wavered in his conviction that there was more to say than quantum theory told. He urged his colleagues not to give up the search for the secrets of "the Old One." But he may have become discouraged. In a letter to a colleague, he wrote: "I have second thoughts. Maybe God is malicious."

In the 1970s, work motivated by EPR showed that Einstein's "spooky interactions" actually do exist. But they're still spooky. They're the subject of our next chapter.

13

Spooky Interactions
Bell's Theorem

. . . thou canst not stir a flower
Without troubling of a star.

—Francis Thompson

Most physicists paid little attention to EPR, or to Bohr's response. It did not matter whether or not quantum mechanics was complete; it *worked*. It never made a wrong prediction, and practical results abounded. Who cared if atoms lacked physical reality before being observed? Working physicists had no time for "merely philosophical" questions.

Shortly after EPR, physicists gave their attention to the Second World War and developed radar, the proximity fuse, and the atom bomb. Then came the politically and socially "straight" 1950s. In physics departments a conforming mind-set increasingly meant that an untenured faculty member might endanger a career by a serious interest in the fundamentals of quantum mechanics. Even today it is best to explore the meaning of quantum mechanics while also working a "day job" on a mainstream physics topic. Since Bell's theorem, however, physicists, especially younger physicists, show increasing interest in that exploration.

Bell's theorem has been called "the most profound discovery in science in the last half of the twentieth century." It rubbed physics' nose in the weirdness of quantum mechanics. As a result of Bell's theorem and the experiments it stimulated, a once "purely philosophical" question has now been answered in the laboratory: There *is* a universal connectedness. Einstein's "spooky interactions" *do* in fact exist. Any objects that have ever interacted continue to instantaneously influence each other. Events at the edge of the galaxy influence what happens at the edge of your garden. Though these effects are completely undetectable in a normally complex situation, they now get attention in industrial laboratories because they may also make possible fantastically powerful computers.

John Stewart Bell

John Bell was born in Belfast in 1928. Though no one in the family had ever had even a secondary school education, his mother promoted learning as the way to the good life, in which you "could wear Sunday suits all week." Her son became an enthusiastic student and, by his own assessment, "not necessarily the smartest but among the top three or four." Eager for knowledge, Bell spent time in the library instead of going off with the other boys, which he would have done had he been, he says, "more gregarious, more socially adequate."

Early on, philosophy attracted Bell. But finding each philosopher contradicted by another, he moved to physics, where "you could reasonably come to conclusions." Bell studied physics at Queen's, the local university. In quantum mechanics, the philosophical aspects interested him most. He felt the courses concentrated too strongly on applications of the theory.

Nevertheless, he finally went to work in an almost engineering role, the design of particle accelerators, eventually at CERN (the European Center for Nuclear Research) in Geneva. But he also gained fame for important work on particle theory. He married a fellow physicist, Mary Ross. Though they worked independently, Bell writes that in looking through his collected papers, "I see her everywhere."

At CERN, Bell concentrated on the mainstream physics that he felt he was paid to do, and of which his colleagues approved. He restrained his interest in the strangeness of quantum mechanics for years, but eventually an opportunity to explore these ideas came in 1964 when he was on sabbatical leave. "Being away from the people who knew me gave me more freedom, so I spent some time on these quantum questions," he tells. The momentous result is what we now call "Bell's theorem." It allows the actual demonstration of aspects of our world that previously could be treated only as philosophical questions.

I (Bruce) shared a taxi and conversation with John Bell in 1989 on the way to a small conference in Erice, Sicily, that focused on his work. At the conference, with wit, and in his Irish voice, he firmly emphasized the depth of the unsolved quantum enigma. In big, bold letters on the blackboard he introduced his famous abbreviation, FAPP, "for all practical purposes," and warned against falling into the FAPPTRAP: accepting a merely FAPP solution. As department chair at the time, I was able to invite Bell to spend some time in our physics department at the University of California, Santa Cruz, and he tentatively agreed. But the next year John Bell suddenly died.

Figure 13.1 John Bell. © Renate Bertlmann.
Courtesy Springer Verlag

Bell's Motivation

Recall that EPR, while accepting the predictions of quantum theory as correct, claimed that the theory's observation-created reality arose from its neglect of an underlying reality, which came to be called "hidden variables." EPR's argument assumed that the behavior of objects could be affected only by physical forces, and any object could otherwise be considered separate from the rest of the world. In particular, two objects could be separated so that the behavior of one could in no way affect the other.

In refuting EPR, Bohr claimed that what happened to one object could indeed "influence" the behavior of the other instantaneously, even though no physical force connected them. Einstein derided Bohr's "influences" as "spooky interactions."

For thirty years, no experimental result could decide between Einstein's hidden variables and Bohr's "influences." Moreover, physicists tacitly accepted a mathematical theorem purporting to show it impossible for a theory with hidden variables to produce the predictions of quantum theory. Since Einstein accepted the predictions of quantum theory as correct, this theorem undercut his argument for hidden variables.

While John Bell enjoyed sabbatical freedom exploring such things, he was surprised to come across a counterexample to the no-hidden-variables proof. He discovered that, twelve years earlier, David Bohm had developed a logically sound interpretation of quantum mechanics that included hidden variables. "I saw the impossible done," said Bell.

After finding where the no-hidden-variables proof went wrong, Bell pondered: Since hidden variables might exist, do they *actually* exist? Is there some observable way in which a world where hidden variables do exist differs from the strange world quantum theory describes, a world where reality is created by observation and objects are connected by mysterious influences? Bell wanted to understand what the quantum calculations he did really meant: "You can ride a bicycle without knowing how it works. . . . In the same way we [ordinarily] do theoretical physics. I want to find the set of instructions to say what we are really doing."

Bell's Theorem

Because EPR's argument did not challenge any of quantum theory's predictions, it could set up no experimental confrontation with the theory. Bell took a different tack. He constructed a "straw man" for experiments to try to knock down. If his straw man survived experimental challenge, quantum theory would be shown fundamentally wrong.

Bell's theorem in a nutshell: Suppose that objects in our world *do* have physically real properties that are not created by observation. And further suppose that two objects can always be separated from each other so that what happens to one cannot affect the other. For short, we'll call these two suppositions "reality" and "separability." From these two premises, both denied by quantum theory, Bell deduced that certain observable quantities had to be larger than other observable quantities. This *experimentally testable* prediction of Bell's theorem is "Bell's inequality." (The most common, experimentally observed quantities here are the rates at which twin-state photons display different polarizations when their polarizers are set at different angles.)

If it is found that Bell's inequality is violated, one or both of the premises that lead to the inequality must be wrong. In other words, if Bell's inequality is violated in actual experiments, our world cannot possibly have both reality and also separability.

All this is pretty abstract. Philosophers and mystics have talked of reality and separability (or its opposite, "universal connectedness") for millennia. Quantum mechanics put those issues squarely in front of us. Bell's theorem allows these ideas to be tested.

In a "reasonable" world, objects should have properties that are real. That is, the properties of an object should not be created by their observation. Moreover, in a reasonable world, objects should be separable. That is, they should affect each other only by physical forces, not by weird, faster-than-light "influences." The Newtonian world described by classical physics is, in this sense, a reasonable one. The world described by quantum theory is not. Bell's theorem allows a test to see whether it is possible that our world is actually reasonable and whether it is possibly only the quantum description that is unreasonable.

We won't go for suspense. When the experiments were done, Bell's inequality was violated. Bell's straw man was knocked down—as he expected it would be. Our world does not have both reality and separability—one, perhaps, but not both. And we immediately admit to not truly understanding what the world being unreal or having a universal connectedness would imply.

We will demonstrate Bell's theorem with twin-state photons. They were the most important actual test. We apply Bell's assumptions of reality and separability to these photons and end up with a Bell inequality. Nick Herbert invented the general idea we use.

To derive Bell's theorem, we start with his premises of reality and separability. We will assume our twin-state photons are reasonable photons in that each one has a real polarization, that its polarization is not observer created. And we will assume that our photons are separable and that an observation of the polarization of one photon does not affect its remote twin. (We are, of course, ascribing properties to twin-state photons that quantum theory denies them.)

To be concrete, we have a specific mechanical picture in mind. However, the logic we use in no way depends on any aspect of this mechanical model except its reality and separability. Bell's mathematical treatment was completely general. It did not even specify photons. It just assumed reality and separability.

Derivation of Bell's Inequality

If you only skim or even skip this rough derivation of Bell's inequality and just accept the result, you will not be much hampered in understanding the rest of the book. You can jump way down to the next section, "The Experimental Tests." But the story we tell here is not that complex, and the result is profound.

An Explicit Model

To show each photon's polarization as graphically real, we show it as a "stick." (figure 13.2). The random angle of the stick is the hidden variable that determines which path a photon takes upon encountering a polarizer.

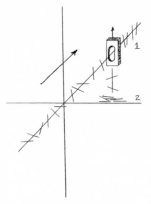

A polarizer in this mechanical model is a plate with an oval opening whose long dimension is the "polarizer axis." A photon whose polarization stick is close to the polarizer axis will pass through the polarizer to go on Path 1. One whose polarization is less close will hit the polarizer to go on Path 2. The important point is that each photon's behavior at the polarizer is determined by a physically real property of that photon, the angle of a solid stick, not something created by observation. This is our version of the reality assumption Bell's proof starts with.

The stick model does not account properly for all the behavior of polarized light. But sticks are simple to visualize and fit our purpose well. You will see that our logic ultimately depends on nothing about the stick model except its reality and separability. The sticks are merely stand-ins for *any* hidden variable.

Figure 13.2 A stick model of photons and a polarizer

We will describe four Alice-and-Bob thought experiments. These experiments are much like the EPR experiment described in our previous chapter. (In fact, Bell's theorem experiments are sometimes loosely referred to as EPR experiments.) But there is a big difference: In the EPR case, Einstein's "hidden variables" and Bohr's "influences" led to the same experimental outcome; the disagreement between Bohr and Einstein was only a difference of interpretation. In the stick model, and in the actual Bell's theorem experiments, the outcome for Einstein's "hidden variables" and Bohr's "influences" will be different.

In each of our four experiments, twin-state stick photons with identical polarizations (identical stick angles) are emitted in opposite directions from a source midway between Alice and Bob. Since the photons fly apart from each other at the speed of light, nothing physical, not even light, can get from one experimenter to the other in the time between the arrivals of the twin-state photons at their respective polarizers. Therefore, we assume that what happens to a photon at one polarizer cannot possibly affect its twin at the other. This is the separability assumption that Bell's theorem starts with. As in the EPR case, Alice and Bob identify two photons as being twins by their identical arrival times and keep track of whether their Path 1 or Path 2 detector recorded each photon.

Experiment I

This first experiment is essentially a repeat of the original EPR experiment, but with our stick photons that have real polarizations and cannot affect each other. Alice and Bob each have their polarizer axes aligned vertically. They record a "1" every time their Path 1 detector records a photon and a "2" every time their Path 2 detector records a photon. They each end up with a long string of random 1s and 2s.

After recording a large number of photons, Alice and Bob come together and compare their results. As in the original EPR experiment, they find their data streams identical. Whenever Alice's photon went on Path 1, its twin, having the

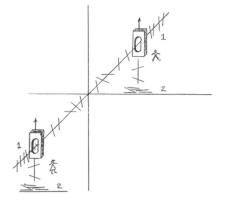

same stick angle, did the same thing at Bob's polarizer. Whenever Alice's went on Path 2, so did Bob's. Not a single difference in behavior is seen. This confirms that their photons were indeed twins.

Alice and Bob find nothing strange in this correlation. Their twin photons indeed had identical polarization, identical stick angles. (In quantum theory, where polarization is observer created, the twin-photon correlation must be explained by a mysterious "influence" instantaneously exerted on a photon by the observation of its twin.)

Experiment II

This is the same as experiment I, except that this time Alice rotates her polarizer by a small angle we'll call Θ. Bob keeps his polarizer axis vertical.

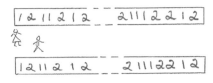

Figure 13.3 Experiment I

Both experimenters take the same kind of data once more. This time some photons that would have gone through Alice's polarizer on Path 1, had she not rotated it, now go on Path 2, and vice versa. We can know what would have happened because the behavior of the photons is determined by their real stick angles, not by Alice's observation. This is our reality assumption. Bob's photons are unaffected by Alice's polarizer rotation or by what happened to their twins at Alice's polarizer—this is our separability assumption.

When Alice and Bob come together this time to compare their data streams,

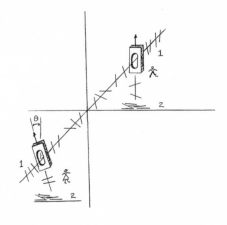

they find some mismatches—we'll call them "errors." When some of Alice's photons went on her Path 2, their twins at Bob's polarizer went on his Path 1, or vice versa. The percentage of errors would be small for small angle, Θ, and would increase for a somewhat larger Θ. (In Experiment I, the percentage of errors was zero for no angle change: Θ = zero.) Let's say, for example, that Alice changed what would have happened for five percent of her photons. She thus caused an error rate of five percent.

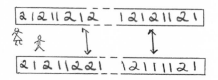

Figure 13.4 Experiment II

Experiment III

This is exactly the same as Experiment II, except that Bob rotates his polarizer by the angle Θ, while Alice returns hers to the vertical. Since the situations are symmetrical, the error rate again would be five percent, assuming that the number of photon pairs recorded was large enough that statistical error was negligible.

Experiment IV

This time both Alice and Bob each rotate their polarizers by the angle Θ. If they each rotated in the same direction, it would amount to no rotation at all; their polarizers would still be aligned. So Alice rotates hers counterclockwise by Θ, while Bob rotates his clockwise by Θ.

Alice, rotating her polarizer by Θ, changes the behavior of her photons by the same amount as in Experiment II. She changes what would have happened to five percent of her photons. The situation is symmetrical. Bob's polarizer rotation by Θ changes the behavior of five percent of his photons from what would have happened.

Since Alice and Bob each changed the behavior of five percent of their photons, and

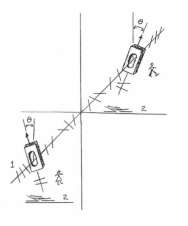

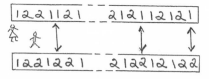

Figure 13.5 Experiment IV

since every change could show up as an error when their data streams are compared, we might expect an error rate of ten percent. There is no way to get a greater error rate with these separable photons in a statistically large sample.

We might, however, get a *smaller* error rate. Here's how: It might be that for some pairs of twin-state photons, both Alice and Bob each caused their twin to change its behavior. The two photons of such twin-state pairs would thus behave identically. The data for such twin-state pairs would not be recorded as errors.

As an example of such a double change of behavior, consider almost vertical twin-state photons that would both go on Path 1 at Alice and Bob's polarizers if their axes were both close to the vertical. If Alice and Bob each rotated their polarizers in opposite directions, as they did in Experiment IV, they could both send this pair of twins on their Path 2. They would not record this double change as an error.

Because of such double changes, when Alice and Bob compare their data streams in Experiment IV, the error rate will likely be less than the five percent error rate Alice alone would cause *plus* the five percent error rate that Bob alone would cause. In Experiment IV the error rate they will see is likely less than ten percent. In a large statistical sample it cannot be greater.

That's it! We've derived a Bell inequality: *The error rate when both polarizers are rotated by* Θ *(in opposite directions) is equal to, or less than, twice the error rate for the rotation by* Θ *of a single polarizer.*

As we said above, our polarization sticks were merely stand-ins for *any* hidden variable within a photon. They merely represented some real property of the photon determining whether it will go on Path 1 or Path 2 for a particular polarizer orientation. We could, for example, have said that each photon is steered by a little "photon pilot" and that a polarizer is just a traffic sign with its orientation being an arrow the pilot looks for. The photon pilot carries a travel document instructing him to steer his photon on Path 1 or Path 2 depending on the angle of the arrow. The hidden variable is now the physically real instruction set printed on the pilot's travel document. Our pilot story emphasizes that the only actual assumptions in our derivation of Bell's inequality were the physical reality of each photon's polarization and the separability of the two twin-state photons, that is, what happened to one twin did not affect the other.

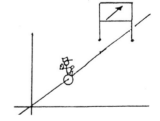

Figure 13.6 The photon pilot

What if experiments with actual twin-state photons found Bell's inequality *not* violated, that it held true? That would be quantum theory's first-ever wrong prediction. But this result would tell us little about reality or separability.

Suppose actual experiments showed Bell's inequality to be violated. That would mean that our world definitely lacks either reality or separability or both. And we will see in what follows that a violation in any one case—twin-state photons, for example—means a lack of reality or separability for everything such photons could possibly interact with. That is *everything*.

Henry Stapp has rederived Bell's inequality without using the reality assumption—this is an important extension of Bell's proof. It means that an experimental violation of Bell's inequality would show that our world definitely lacks separability. It would leave reality as an open question. (Quantum theory, of course, would still say reality was created by observation.)

The Experimental Tests

In 1965, when Bell's theorem was published, it was a mild heresy for a physicist to question quantum theory or even to doubt that the Copenhagen interpretation settled all the philosophical issues. As a physics graduate student at Columbia University, John Clauser was nevertheless intrigued. Though Bell's inequality was not in a form that could be tested experimentally (and Bell thought it would be many years before that would happen), Clauser figured out a way to do it.

Off to Berkeley as a postdoc to work on radio astronomy with Charles Townes, Clauser presented his idea for a test of Bell's inequality. Townes released him from his commitment to work on astronomy, even continued his financial support, and encouraged him to go for it—in spite of some other faculty considering the project pointless, even silly. With equipment loaned him by another Berkeley faculty member, Clauser and a graduate student did a brilliant experiment with twin-state photons.

They found Bell's inequality was *violated!* Moreover, it was violated in just the way quantum theory predicts. (To avoid a common misstatement, we emphasize that Bell's *inequality* was violated. Bell's theorem, the mathematical proof deriving the inequality, is, of course, a logical result not testable by experiment.)

A Bottom Line for the Experimental Results

Clauser's experiments ruled out, in physics terminology, "local reality" or "local hidden variables." The experiments showed that the properties of objects in our world have an observation-created reality *or* that there exists a universal connectedness, or both. In these experiments, quantum theory survived its most serious challenge in decades.

Clauser writes: "My own . . . vain hopes of overthrowing quantum mechanics were shattered by the data." Confirming quantum theory's predicted violation of

Bell's inequality, he showed instead that a "reasonable" description of our world, that is, a description with separability and reality, would never be possible.

We can never be certain that a particular scientific theory is correct. Some day a better theory might supersede quantum theory. But we now know that this better theory must also describe a world without separability. Before Clauser's result, we could not know this.

Unfortunately for Clauser, in the early 1970s, investigation of the fundamentals of quantum mechanics was not yet considered proper physics in most places. When he sought an academic position (including an opening in our own department at the University of California, Santa Cruz), his work was met with scorn. "What has he done besides checking quantum theory? We all *know* it's right!" was a typical misunderstanding of Clauser's accomplishment. Clauser got a job in physics, but not one in which he could participate in the wide-ranging investigations he launched.

A decade later, with advanced technology and a more receptive atmosphere for exploring quantum fundamentals, Alain Aspect in Paris duplicated Clauser's results with far greater accuracy, showing that the violation of Bell's inequality was by precisely the amount predicted by quantum theory. His faster electronics established that no physical effect could possibly propagate from one polarizer to another in time for the observation of one photon to physically affect the other. This closed a small loophole in the experiment of Clauser, whose electronics were not quite as fast. If John Bell had not died, Bell, Clauser, and Aspect might well share a Nobel Prize.

The Aspect result will not be the end of the story. In Bell's words:

It is a very important experiment, and perhaps it marks the point where one should stop and think for a time, but I certainly hope it is not the end. I think the probing of what quantum mechanics means must continue, and in fact will continue, whether we agree or not that it is worth while, because many people are sufficiently fascinated and perturbed by this that it will go on.

Where Does the Violation of Bell's Inequality Leave Us?

Separability First

"Separability" has been our shorthand term for objects being affected only by physical forces. These objects' behavior is not otherwise influenced by what happens to the things they once interacted with. The experiments show that a single twin-state photon's behavior is instantaneously connected to that of its twin. That connectedness can extend beyond the photon pair to macroscopic things.

Consider Schrödinger's "hellish contraption." Suppose a twin-state photon entering the cat box would trigger a cyanide release if that photon displayed vertical polarization and would not trigger the release if it displayed horizontal polarization. The fate of the cat would be (randomly) determined by the remote observation of the polarization of the photon's twin. Moreover, the cat and that remote observer are then forevermore entangled. The lack of separability of any two objects, twin-state photons, for example, establishes the lack of separability generally.

We talk in terms of twin-state photons because that situation is readily described and subject to experimental test. In principle, however, any two objects that have ever interacted are forever entangled. The behavior of one instantaneously influences the other. An entanglement exists even if the interaction is through each of the objects having interacted with a third object. In principle, our world has a universal connectedness.

Quantum entanglement for large objects is generally too complex to notice. But not always. Quantum entanglements of the essentially macroscopic elements of future quantum computers are intended to make use of this connectedness.

Reality

"Reality" has been our shorthand term for objects having physically real properties that are not created by observation. If the polarization of a photon is not a physical reality until it is observed, neither, for example, is the living or dead state of Schrödinger's cat entangled with that photon. Quantum theory has no boundary between the microscopic and the macroscopic.

We might stretch our minds to envision a nonseparable world, a universal connectivity. Conceiving a nonreal world is more difficult. It would be a world where what we call physical reality is created by observation. Would that be *conscious* observation? We postpone this question until after we address consciousness itself.

Induction

We now note that Bell's theorem, in addition to reality and separability, assumes the validity of inductive reasoning. "All crows we have seen are black; therefore, all crows are black," is reasoning by induction—it is going from particular cases to a generality. Our crow example assumes that the already seen crows are representative of all crows. Strictly speaking, it is possible that every not-yet-seen crow is green. Reasoning by induction has logical problems. But all science is based on induction.

For twin-state photons, both in our model and in the actual experiments, the induction assumption implies that the stream of photons for a particular polarizer angle is representative of all the photons in the experiment. For example, we assumed that Alice and Bob could have chosen to do Experiment IV with the photon stream with which they *actually* did Experiment II, and that if they had done so, they would have gotten essentially the same results. Denying this assumes a conspiratorial world that seems even stranger than the world of quantum theory.

Is It Einstein for Whom the Bell Tolls?

Both Einstein and Bohr died before Bell presented his theorem. We are sure Bohr would have predicted the experimental result confirming quantum theory. It is not clear what Einstein would have predicted had he seen Bell's proof. He said he believed that quantum theory's predictions would always be correct. But how would he feel if the predicted result was an actual demonstration of what he called "spooky interactions"?

In the EPR argument, Bell, Clauser, and Aspect showed Bohr to be right and Einstein wrong. But Einstein was right that there was something to be troubled about. It was Einstein who brought quantum theory's full weirdness up front. It was his objections that stimulated Bell's work and that continue to resonate in today's attempts to come to terms with the strange worldview quantum mechanics forces on us.

According to Bell:

In his arguments with Bohr, Einstein was wrong in all the details. Bohr understood the actual manipulation of quantum mechanics much better than Einstein. But still, in his philosophy of physics and his idea of what it is all about and what we are doing and should do, Einstein seems to be absolutely admirable. . . . . [T]here is no doubt that he is, for me, the model of how one should think about physics.

Does Quantum Mechanics Support Mysticism

Some imply that the sages of ancient religions intuited aspects of contemporary quantum mechanics. The argument can go on to claim that quantum mechanics provides evidence for the validity of such religions. Such arguments are not compelling. The mysteries of reality and connectedness have actually loomed large in various philosophies for many centuries.

The Newtonian worldview can be seen as dismissing such ideas. Quantum mechanics tells us the mysteries still exist. In this most general sense, one can argue that the findings of physics do support the thinking of ancient sages. (When Bohr was knighted, he put the Yin-Yang symbol in his coat of arms.)

Quantum mechanics tells us strange things about our world, things that we do not fully comprehend. This strangeness has implications beyond what we generally consider physics. We might therefore be tolerant when nonphysicists incorporate quantum ideas into their own thinking—even if they do so with less than complete understanding, or even a bit incorrectly.

We are, however, disturbed, and sometimes embarrassed, by cavalier, perhaps intentional, misuse of quantum ideas, as a basis for certain medical or psychological therapies (or investing schemes!), for example. A touchstone test for such misuse is the presentation of these ideas with the implication that the notions promulgated are derived from quantum physics rather than merely suggested by it.

Quantum mechanics does, however, provide good jumping-off points for imaginative stories. The teleportation in *Star Trek*—"beam me up, Scotty"—is a wild, but acceptable and imaginative, extrapolation of the transmission of quantum influences displayed in EPR-type experiments. Such stories are fine, if it is clear, as in *Star Trek*, that they are fiction. Unfortunately, that is not always so.

Extrasensory perception (ESP) and other conceivable "paraphenomena" warrant special mention and get it in chapter 16.

The universal connectedness predicted by quantum theory ("thou canst not stir a flower / Without troubling of a star") is now demonstrated. It supports wild speculations. Some of those speculations are alternatives to the Copenhagen interpretation. We treat them in our next chapter.

14

What's Going On?
Interpreting the Quantum Enigma

You know something's happening here, but you don't know what it is.

—Bob Dylan

Every interpretation of quantum mechanics involves consciousness.

—Euan Squires

Physicists and Consciousness

To their dying days, Bohr and Einstein disagreed about quantum theory. For Bohr, the theory with its Copenhagen interpretation was the proper basis for physics. Einstein rejected Copenhagen's concept of a physical reality created by observation. He nevertheless accepted the *goal* of the Copenhagen interpretation, which was to allow physics to move on without dealing with consciousness. Most physicists would vehemently agree that consciousness itself is not within the physics discipline, not something to be studied in a physics department.

It is not that physicists are averse to ranging widely. For example, a famous mathematical treatment of predator–prey relations (foxes and rabbits isolated on an island) was published in *Reviews of Modern Physics* in 1971. On Wall Street, physicists model arbitrage (and are called "quants"). Even one of us (Bruce) has strayed into biology to analyze how animals detect Earth's magnetic field. Such things are happily accepted as part of the physics discipline, while the study of consciousness is not. Here is a reasonable working definition of physics that makes that attitude understandable: Physics is the study of those phenomena that are successfully treatable with well-specified and testable models.

For example, physics treats atoms and simple molecules. Chemistry, on the other hand, deals with all molecules, most of whose electron distributions cannot be well specified. A physicist might study a readily characterized biological system, but the functioning of a complex organism lies in the domain of biologists.

Anything not successfully treatable with a well-specified and testable model is rather quickly defined out of physics. When we focus on consciousness in the following chapter, we offer no well-specified, testable model—no one has ever come up with one. Until such a model is developed, consciousness won't qualify for study as physics.

This may be reason enough for not studying consciousness in physics departments, but it hardly explains the emotion that talking of our discipline's *encounter* with consciousness can arouse. Recently, I (Fred) gave a talk in our physics department reporting on two conferences I attended. At one, honoring Princeton University quantum cosmologist John Wheeler on his ninetieth birthday, talks on cosmology and the fundamentals of quantum mechanics referred to consciousness. The other conference was "Quantum Mind 2003," sponsored by the University of Arizona.

When I spoke of our interest in the issues of consciousness raised at the two conferences, I was heckled by some senior faculty: "You guys are taking physics back to the Dark Ages!" And: "Spend your time doing good physics, not this nonsense!"

Physics graduate students in the audience, on the other hand, seemed fascinated. Not surprisingly. Younger physicists today are generally more open to the idea that there are problems with the foundations of quantum theory.

We're not unsympathetic with the reaction of some colleagues. Our discipline's encounter with consciousness sometimes embarrasses us as well—particularly when the encounter is claimed to confirm metaphysical philosophies. Even though, as we've said earlier, quantum mechanics can seem to resonate with such ideas.

Classical physics, with its mechanical picture of the world, has been taken to deny almost all metaphysics. Quantum physics denies that denial: It hints at the existence of something beyond what we usually consider physics—beyond what we usually consider the "physical world." *But that's the extent of it!* Physics can certainly suggest directions for speculation. We should, however, be careful—in dealing with the mysteries of quantum mechanics, we walk the edge of a slippery slope.

A recent movie provides a good illustration of the problem. It's strangely titled *What the #$*! Do We (K)now!?* (It's informally called *What the Bleep Do We Know?*) *Time* magazine describes it as "an odd hybrid of science documentary and spiritual revelation featuring a Greek chorus of Ph.D.s and mystics talking about quantum physics." The movie uses special effects to display quantum phenomena with macroscopic objects—for example, exaggerating the uncertainty of the position of a basketball. That's legitimate and is understood as hyperbole. The movie's allusion to mysteries in quantum mechanics being connected to the realm of consciousness is also valid. But then the movie blends over to "spiritual revelations" and to the implication of a quantum connection with the channeling of a 35,000-year-old Atlantis god named Ramtha and other such nonsense.

What's in the minds of the audience leaving the theater? If it's that physicists who use quantum mechanics spend their time dealing with the "spiritual revelations" the movie describes, we're embarrassed. If viewers think the physicists in the movie expressing these mystical ideas represent more than the very tiniest fraction of the physics community, they've been misled. The movie slides far down the slippery slope.

The antidote for sensationalistic, misleading treatments of the implications of quantum mechanics would be for the physics discipline to be more open to some discussion of the quantum enigma in introductory physics courses, for example. We should not try to keep our skeleton in the closet.

Why Interpretations?

Quantum mechanics forces us to accept that the mechanistic Newtonian view of the world—and the intuitions fostered by it—are fundamentally flawed. The Copenhagen interpretation provides a sound logical way for physicists to ignore the quantum weirdness and get on with the usual business of physics. Appropriately, most physicists do just that. But it is also fascinating to explore what Nature seems to be telling us.

As John Bell says:

> Is it not good to know what follows from what, even if it is not necessary FAPP ["for all practical purposes"]? Suppose for example that quantum mechanics were found to *resist* precise formulation. Suppose that when formulation beyond FAPP is attempted, we find an unmovable finger obstinately pointing outside the subject, to the mind of the observer, to the Hindu scriptures, to God, or even only Gravitation? Would that not be very, very interesting?

Going beyond FAPP to interpret quantum theory is today a growth industry, and a contentious field. Of course, only a tiny fraction of all physicists are involved. Of the several interpretations in current discussion, each looks in a different way at what quantum mechanics tells us about our world—and in some ways about us. At times, different interpretations seem to say the same thing in different terms. On the other hand, two interpretations might even contradict each other. That's okay—while scientific theories must be testable, interpretations need not be. Some, though, intriguingly hint at testable results.

There is no way to interpret quantum theory without in some way addressing consciousness. Most interpretations accept the encounter but offer a rationale for avoiding a relationship. They usually start with the presumption that the physical world should be dealt with independently of the human observer. Murray Gell-Mann, for example, begins a popular treatment of quantum physics by saying: "[T]he universe presumably couldn't care less whether human beings evolved on some obscure planet to study its history; it goes on obeying the quantum-mechanical laws of physics irrespective of observation by physicists." In talking about classical physics, Gell-Mann's presumption would go *without* saying.

Each interpretation we will discuss is currently defended as the best way to view what quantum mechanics is telling us. Each, however, presents a weird view of the world. How could it be otherwise? We saw the weirdness of quantum mechanics right up front in the most basic experimental findings. Any interpretation explaining those findings that goes beyond FAPP (or "Shut up and calculate!"), must be weird.

Though the interpretations we discuss have been developed with extensive mathematical and logical analysis, we package each in a few nontechnical paragraphs. Thoroughness in understanding them is not crucial for what follows. It's enough to get the flavor of the wide range of views expressed and to see that quantum physics shows that profound questions about our world are wide open. Notice particularly how each interpretation involves consciousness or tries to evade the encounter.

Nine Current Interpretations

Copenhagen

The Copenhagen interpretation, physics' orthodox stance, is the way we teach and use quantum theory. We say little about it here since we devoted a whole chapter to it earlier. In the standard version of Copenhagen, observation creates the physical reality of the microscopic world, but the "observer" can, for all practical purposes, be considered to be the macroscopic measuring device, a Geiger counter, for example.

Copenhagen addresses the quantum enigma by telling us to pragmatically use quantum physics for the microworld and to use classical physics for the macroworld. Since we supposedly never see the microworld "directly," we can just ignore its weirdness, and thus ignore physics' encounter with consciousness. However, as quantum weirdness is seen with larger and larger objects, the ignoring gets harder, and other interpretations proliferate.

Extreme Copenhagen

Aage Bohr (a son of Niels Bohr, and also a Nobel laureate in physics) and Ole Ulfbeck hold that the usual Copenhagen interpretation does not go far enough. Where a standard Copenhagen interpretation allows physics to ignore its encounter with consciousness by confining observer-created reality to the microscopic quantum world, they explicitly deny the existence of the microworld. In this view, there are no atoms.

Bohr and Ulfbeck intend their outlook to apply generally but discuss it in terms of the clicks of a Geiger counter and the changes in a piece of uranium, whose radioactivity causes counters to click.

We normally consider uranium nuclei to randomly emit alpha particles (helium nuclei) and then become thorium nuclei. An alpha may be a widely extended wavefunction impinging on a Geiger counter. In the standard Copenhagen interpretation, the extended wavefunction of the alpha particle is, for all practical purposes, collapsed by the Geiger counter to the position at which the counter observed it.

Bohr and Ulfbeck find such a for-all-practical-purposes resolution unacceptable. Taking the bull by the horns, they claim that atomic-scale objects do not exist at all. Nothing moved through the space between the piece of changed uranium and the clicking Geiger counter. Clicks in counters are "genuinely fortuitous" events that are correlated with changes in a remote piece of uranium without the intermediary of alpha particles.

As they say it:

> The notion of particles as objects in space, taken over from classical physics, is thereby eliminated. . . . The click being genuinely fortuitous, is no longer produced by a particle entering the counter, as has been a foregone conclusion in quantum mechanics. . . . The downward path from macroscopic events in spacetime, which in standard quantum mechanics continues into the region of particles, does not extend beyond the onset of clicks.

Accordingly, when chemists, biologists, and engineers talk of photons, electrons, atoms, and molecules, they are presumably speaking of mathematical objects without physical reality. No photons pass through the space between the light bulb and your eye. No air molecules bounce off the canvas sheet to push the sailboat through the water. This interpretation shows how far some physicists will go to evade the encounter with consciousness.

Decoherence and Consistent Histories

Some years ago, a physicist would likely use the word "collapse" to describe the process of observation in which a superposition state wavefunction changes to an observed single reality. Instead of "collapse," a physicist today might use the word "decoherence." It refers to a now well-studied process in which the wavefunction of a microscopic object interacts with the macroscopic environment to produce the result the Copenhagen interpretation describes as collapse.

Let's go to our box-pair example. Consider an atom whose wavefunction is simultaneously in two boxes. We now send a photon through tiny holes in one of the boxes. Were the atom in that box, the photon would bounce off the atom in a new direction. Were the atom in the other box, it would go straight through unchanged in direction. Since the atom is actually simultaneously in both boxes, the photon does both things. The atom's wavefunction becomes entangled with that of the photon. The parts of the atom's wavefunction in each box would no longer by themselves have a coherent phase relationship; they would have "decohered."

If the photon does not interact with anything else, a tricky two-body interference experiment with a set of box pairs and photons could demonstrate that the atom was still simultaneously in both boxes—and that the photon had both bounced off the atom and gone through an empty box.

Suppose that the photon passes through our boxes and then hits a solid body and interacts with many other atoms. Assuming some thermal randomness, it is possible to calculate the fantastically short time after which an interference experiment becomes essentially impossible. One therefore cannot display a quantum enigma. We are left with a classical-like probability for the atom actually existing in either box of its pair. We have the perception of a unique reality. Since no observer, conscious or otherwise, need be mentioned, some consider this to resolve the observer problem.

But those classical-*like* probabilities are still probabilities of what will be observed, not true classical probabilities of what supposedly actually exists. Consciousness is still encountered. For example, W. H. Zurek, a developer of this interpretation, writes in his major treatment of decoherence:

An exhaustive answer to this question [the perception of a unique reality] would undoubtedly have to involve a model of "consciousness," since what we are really asking concerns our (observer's) impression that "we are conscious" of just one of the alternatives.

"Consistent histories" can be seen as an extension of the decoherence idea. (It is sometimes called "decoherent histories.") This interpretation is boldly presented in order to apply quantum theory to the entire universe, from beginning to end. No observers were around in the early universe, and no *external* observers exist at any time; the universe includes everything. Since one can't deal with the infinite complexity of the universe, one treats only certain aspects and averages over the rest.

For a very rough idea of how this can work, consider our atom on its way to the box pair passing through a thin gas of very light atoms. Gently bouncing along, it is not strongly deviating from its path, but the parts of the wavefunction on each path, changing phase a tiny bit with each bounce, decohere enough that no interference experiment is possible, for all practical purposes. By averaging over the vast number of histories, one for every possible series of bounces, we come to two course-grained histories, one for the atom in each box. Now we claim that only one of these two histories is an actual history, and the other is just a history that was possible.

In their development of this interpretation, Gell-Mann and James Hartle discuss the evolution of an IGUS, an "information gathering and utilizing system." The IGUS can eventually become an observer who has at least the illusion of conscious free will. We again see no escape from the encounter with consciousness.

Many Worlds

The many-worlds interpretation accepts literally what quantum theory says. Where the Copenhagen interpretation has observation collapsing the atom's wavefunction into a single box—and Schrödinger's cat into the living *or* dead state—the many-worlds interpretation "just says 'no'" to collapse. If quantum theory says the cat is simultaneously alive and dead, so be it! In one world, Schrödinger's cat is alive, and in another it is dead. "Many worlds" may be the most bizarre description of reality ever proposed.

Hugh Everett came up with the idea in the 1950s to allow cosmology to treat a wavefunction for the universe. Since there are no "observers" external to the entire universe, the many-worlds interpretation resolves the mystery of the conscious observer by the sensible-seeming ploy of including consciousness as part of the physical universe described by quantum mechanics.

In the Copenhagen interpretation, choosing an interference experiment, you can prove that an atom's wavefunction (the atom itself) was in a superposition state spread over two boxes. However, if you look in a box, you'll find the whole atom to be wholly in a single box. You can choose to demonstrate either of two contradictory situations.

In the many-worlds interpretation, when you look into one of the boxes, you entangle with the atom's superposition state. You go into a superposition state *both* of having seen the atom in the box you looked in and also of having seen that box empty. There are now two of you, one in each of two parallel worlds. The consciousness of each one of you is unaware of the other you. Nothing we actually experience conflicts with this bizarre view.

Instead of the look-in-a-box experiment, you might have chosen to do an interference experiment. It is your exercise of free will—your being able to freely choose to do either experiment—that brings about physics' encounter with consciousness. In the many-worlds interpretation, you are part of the universal wavefunction. Everything that can possibly happen as the wavefunction evolves does happen. You both looked in a box and did an interference experiment. You took both options. You exercised no free will.

To bring more than one observer into the picture, let's go back to Schrödinger's cat. Alice looks in the box while Bob is far away. The world splits in two. In one world Alice, call her Alice$_1$, sees a live cat. In the other, Alice$_2$ sees a dead cat. At this point Bob is also in both worlds, but Bob$_1$ and Bob$_2$ are essentially identical. Should Bob$_1$ meet Alice$_1$, he would help her get milk for the hungry cat. Bob$_2$ would help Alice$_2$ bury the dead cat. Macroscopic objects Alice$_2$ and Bob$_1$ exist in different worlds and, for all practical purposes, never encounter each other.

After Bell's theorem and the experiments it allowed, we know we cannot have both reality and separability. In the many-worlds interpretation, there is no separability. And there is no single reality, which is essentially equivalent to no reality.

The many-worlds interpretation stirs strong feelings. One academic author decries it as "profligate" and refers to its proposer as a "chain-smoking, horned-Cadillac-driving, multimillionaire weapons research analyst." (At the time Everett proposed it, he was just a graduate student.) On the other hand, a leader in quantum computing writes that the many-worlds interpretation "makes more sense in so many ways than any previous world-view, and certainly more than the cynical pragmatism which too often nowadays serves as a surrogate for a world-view among scientists." (By "cynical pragmatism" he surely means the unquestioning acceptance of Copenhagen.)

There's an unresolved problem with many-worlds: What constitutes an observation? When does the world split? The splitting into a finite number of worlds is presumably just a way of speaking. Are infinitely many worlds continuously created?

In any event, this interpretation vastly extends what Copernicus started. Not only are we removed from the center of the cosmos to a tiny spot in a limitless universe, but the world we experience is just a minute fraction of all worlds. However, "we" exist in many of them. Though bizarre, the many-worlds interpretation is a fascinating base for speculation.

Transactional

The transactional interpretation approaches the intuitive challenges posed by Schrödinger cats and universal connectedness by allowing the wavefunction to evolve backward as well as forward in time. The future thus affects the past. This does, of course, alter the way we look at what's happening.

For instance, here's an example offered by the transactional interpretation's proposer, John Cramer:

> When we stand in the dark and look at a star a hundred light years away, not only have the retarded light waves from the star been traveling for a hundred years to reach our eyes, but the advanced waves generated by absorption processes within our eyes have reached a hundred years into the past, completing the transaction that permitted the star to shine in our direction.

Reading this, it seems that the backward-in-time approach very much involves an encounter with the conscious observer. But we do end up with the quantum enigma packaged into what appears to be a single mystery.

Bohm

In 1952 a maverick young physicist, David Bohm, did the "impossible" by producing a counterexample to the long-accepted no-hidden-variables proof. He thus showed that quantum theory was not inconsistent with the existence of real particles each with a real position and a real velocity. (Bohm was also a maverick politically. After he refused to testify before the House Un-American Activities Committee, Princeton University fired him, and he could not get another academic job in the United States.) It was Bohm's work that inspired John Bell to challenge the no-hidden-variables proof and eventually produce Bell's theorem.

Bohm starts his interpretation by assuming that his particles, on average, reproduce all the results demanded by the Schrödinger equation. Then, with straightforward mathematics, he deduces a "quantum force" (or "quantum potential") that acts on his particles to make them do just that.

The quantum force guides rather than pushes. Bohm uses the analogy of the radio beacon directing a ship. The universal connectedness intrinsic to quantum theory appears right up front in this interpretation. The quantum force on an object depends instantaneously on the positions of all other objects the one in question has ever interacted with, and with all objects that had ever interacted with those objects—in essence, with everything in the universe.

The Bohm interpretation describes a physically real, completely deterministic world. Quantum randomness appears only because we cannot know the precise initial position and velocity of each particle. There is no unexplained wavefunction collapse, as there is in the Copenhagen interpretation; there is no unexplained splitting of worlds as there is in the many-worlds interpretation. Some claim the Bohm interpretation resolves the observer problem of quantum mechanics, or at least makes it a benign problem, as it is in Newtonian physics.

Others see it differently. Unlike a Newtonian atom that enters a single box of a pair, a Bohmian atom entering a single box also "knows" the position of the other box through the quantum potential. Since the macroscopic box pair and its mirror arrangement are in contact with the rest of the world, the atom is also in contact with the macroscopic device that earlier released the atom. The quantum potential connects all of this from the beginning and even determines where the atom would land in an eventual interference pattern. The human who arranged the box pairs also influences the quantum potential. There is no physical world "out there" separate from the observer in the Bohm interpretation's undivided universe.

As in many-worlds, since there is no collapse, the part of the wavefunction corresponding to what was not observed continues on forever: We may find Schrödinger's cat alive, but the part of the wavefunction containing the possibility of the dead cat—and its owner burying it—goes on. We may ignore this part of the wavefunction for all practical purposes, but in this interpretation it is real and, in principle, has future consequences.

The Bohm interpretation seems to conflict with special relativity, but we do not see this as an insurmountable problem. Bohm himself did not believe his interpretation avoids physics' encounter with consciousness. In their highly technical 1993 book on quantum theory, *The Undivided Universe,* whose title emphasizes the universal connectedness and the nonseparability of the microscopic from the macroscopic, Bohm and Basil Hiley write:

> Throughout this book it has been our position that the quantum theory itself can be understood without bringing in consciousness and that *as far as research in physics is concerned, at least in the present general period,* this is probably the best approach. However, the intuition

that consciousness and quantum theory are in some sense related seems to be a good one, and for this reason we feel it is appropriate to include in this book a discussion of what this relationship might be. (emphasis added)

GRW

To explain why big things appear solidly in one place while atomic-scale objects can be spread-out waves, Ghirardi, Rimini, and Weber—GRW—modify the Schrödinger equation to make wavefunctions randomly collapse every now and then. For things as small as atoms, a collapse occurs only every hundred million years or so.

Such infrequent collapse would not affect an interference experiment taking place in a much shorter time. But were an atom correlated with other atoms as part of a larger object, say, Schrödinger's cat, the atom's collapse to a place characteristic of a living or dead cat would trigger the collapse of the whole cat to the living or dead state. There are so many atoms in a cat that even if a single atom collapsed only every hundred million years, at least one atom would collapse every micromicrosecond. The cat could thus remain in a superposition of living and dead states only briefly.

Strictly speaking, the GRW scheme is not an interpretation of the theory since it proposes a *change* of the theory. There is no experimental evidence for the GRW phenomenon, and the experimental window of possibilities for it narrows. In any event, it would leave the wave–particle problem of microscopic objects and the experimentally confirmed lack of separability in our world still an enigma.

Ithaca

David Mermin of Cornell University in Ithaca, New York, proposing what he calls the "Ithaca interpretation," identifies two "major puzzles": objective probability, which arises only in quantum theory, and the phenomenon of consciousness.

Classical probability is subjective, a measure of an individual's ignorance. *Quantum* probability is objective—the same for everyone. For the atom in a box pair, the quantum probability is not the probability of what *is* but of what anyone would *observe* in a particular experiment. Ithaca takes objective probability as a primitive concept incapable of further reduction and reduces the mysteries of quantum mechanics to this single puzzle.

According to Ithaca, quantum mechanics is trying to tell us that "correlations have physical reality; that which they correlate do not." For example, unobserved twin-state photons have no particular polarization, but they have the same polarization. Only the correlation of their polarizations is a physical reality; the polarizations themselves are not.

What if, for example, we observe a photon with a macroscopic apparatus whose scale reads differently for two superposition states of the photon? If we consider the apparatus quantum mechanically, it merely becomes correlated with the photon. If so, the scale should read both ways. But we always see it read one way or the other.

Here's how Mermin deals with this in the Ithaca interpretation:

> When *I* look at the scale of the apparatus *I know* what it reads. Those absurdly delicate, hopelessly inaccessible, global system correlations *obviously* vanish completely when they connect up with *me*. Whether this is because consciousness is beyond the range of phenomena that quantum mechanics is capable of dealing with, or because it has infinitely many degrees of freedom or special super-selection rules of its own, I would not presume to guess. But this is a puzzle about consciousness which should not get mixed up with efforts to understand quantum mechanics as a theory of subsystem correlations in the nonconscious world. (emphasis original)

Ithaca steps aside from physics' encounter with consciousness to confine the quantum enigma to the problem of objective probability. The encounter with consciousness is not denied. Ithaca assigns consciousness to a "reality" larger than the "physical reality" to which physics, for the present at least, should be restricted.

Quantum Logic

By your choice of experiment, you seem able to logically prove contradictory things. Instead of looking for an explanation of this, you can change the rules of logic to fit the observed facts. That's the approach of quantum logic. Few find this a satisfying solution to the quantum enigma. Can't any conceivable observations be "explained" by adopting rules of logic to fit? It may be an intriguing intellectual exercise and may be useful for analyzing quantum computers, but quantum logic seems to offer little insight into what Nature is trying to tell us.

Two proposals, one by Roger Penrose and another by Henry Stapp, might be called interpretations but actually include physical speculations involving consciousness. We address these in chapter 16.

Some interpretations of quantum mechanics indeed resolve the measurement problem for all practical purposes, but there never *was* a problem for all practical purposes. The predictions of the theory work perfectly—it's the strange worldview that we want to make sense of.

But that's not a goal for most of the physics community. Even though in almost any discussion of the fundamentals of quantum mechanics, the role of the observer—or consciousness—arises, most physicists shun the C-word. We like the way Wheeler puts the dichotomy:

> Useful as it is under everyday circumstances to say that the world exists "out there" independent of us, that view can no longer be upheld. There is a strange sense in which this is a "participatory universe."

Immediately after stating that, Wheeler cautions:

> "Consciousness" has nothing whatsoever to do with the quantum process. We are dealing with an event that makes itself known by an irreversible act of amplification, by an indelible record, an act of registration. . . . [*Meaning*] is a separate part of the story, important but not to be confused with "quantum phenomenon."

We take this as an injunction to physicists (as physicists) to study only the quantum phenomena, not the *meaning* of the phenomena. But some of us, as physicists, or just as wonderers, want to ponder the meaning, to try to understand what's going on. This has long been an attitude of many eminent physicists (including, at times, John Wheeler). It's an attitude that today is gaining acceptance.

That growing acceptance of seeking meaning stimulates challenges to that acceptance. And over-the-top treatments of quantum mechanics—like the movie *What the Bleep?*—make physicists squirm and motivate them to minimize the enigma. And, of course, we keep the skeleton in the closet, and sometimes even deny its existence.

For example, in 1998, an article entitled "Quantum Theory without Observers," spanning two issues of *Physics Today,* argued that several interpretations, principally the Bohm interpretation, eliminate a role for the observer in quantum mechanics. (Bohm himself, quoted above, would not agree.) When such arguments are put forth, it is usually unclear whether the elimination of the observer is supposedly in principle or just for all practical purposes (a FAPPTRAP, to use Bell's put-down of the for-all-practical-purposes argument when used to address fundamental problems). But we emphasize that the attitude of this *Physics Today* article matches the sympathies of the majority of the physics community.

What is certainly true is that eight decades after the Schrödinger equation, the meaning of physics' encounter with consciousness is still in contention. When experts can't agree, you can choose your expert—or speculate on your own.

"What's going on?" is still an open question. "You know something's happening here, but you don't know what it is." Physics has encountered something beyond the realm of "ordinary" physics. And, "Every interpretation of quantum mechanics involves consciousness."

Starting with quantum mechanics, we have encountered consciousness. Our next chapter starts with consciousness and approaches the encounter from the other direction.

The Mystery of Consciousness

*What is meant by consciousness we need not discuss;
it is beyond all doubt.*

—Sigmund Freud

*Consciousness poses the most baffling problems in the
science of the mind. There is nothing that we know
more intimately than conscious experience, but there
is nothing that is harder to explain.*

—David Chalmers

When in our book we discuss the demonstrated quantum facts and quantum *theory* (as distinct from its contending interpretations), we describe the generally accepted position of the physics community. We cannot describe such a consensus in our discussion of consciousness—there is none. Even diametrically opposed positions are strongly held. We have our own take, but, you may notice, we waver.

Until the 1960s, behaviorist-dominated psychology avoided the term "consciousness" in any discussion that presumed to be scientific. What caused the explosion of interest in consciousness in the past couple of decades?

Some attribute it to the striking developments in brain imaging technology that allowed seeing which parts of the brain became active with particular stimuli. According to an editor of the *Journal of Consciousness Studies*:

> It is more likely that the re-emergence of consciousness studies occurred
> for sociological reasons: The students of the 1960s, who enjoyed a rich
> extra-curricular approach to "consciousness studies" (even if some of
> them didn't inhale), are now running the science departments.

Interest in the foundations of quantum mechanics and the connection with consciousness has emerged at the same time as consciousness studies. There's something in the air.

What Is Consciousness?

We have talked about consciousness but never defined it. Dictionary definitions of "consciousness" are little better than those for "physics." We've been using "consciousness" as roughly equivalent to "awareness," or perhaps the *feeling* of awareness. (As Humpty Dumpty told Alice: "When *I* use a word . . . it means just what I choose it to mean," and the philosopher Wittgenstein would more or less agree.) It's often pointed out that we can know of the existence of consciousness in *no* other way than through our first-person feeling of awareness or the second-person reports of others. (In the following chapter we suggest a quantum challenge to this limitation.)

We do not include in our discussion many of the things found in discussions of consciousness from a psychological point of view. For example, we will not talk of optical illusions, mental disturbances, self-consciousness, or Freud's seat of hidden emotions, the *unconscious*.

Our concern is with the consciousness central to the quantum enigma—the awareness that appears to affect physical phenomena. Our simple example was that your observation of an object wholly in a single box *caused* it to be there, because you presumably *could* have chosen to cause an interference pattern establishing a contradictory situation, whereby the object would have been a wave simultaneously in two boxes.

Does such a demonstration necessarily require a *conscious* observer? Couldn't a not-conscious robot, or even a Geiger counter, do the observing? That most commonly voiced objection to consciousness being required comes up—and is refuted—in our next chapter. For now, just recall that, according to quantum theory, if that robot or Geiger counter were not in contact with the rest of the world, it would merely entangle to become *part* of a total superposition state—as did Schrödinger's cat. In that sense it would not truly *observe*.

Our box-pair demonstration of the involvement of consciousness rests on the assumption that we *could* have chosen to do an experiment other than the one we actually did, that we have free will. The same is true for our Bell-type experiments demonstrating Einstein's "spooky interactions." The existence of a quantum enigma depends crucially on free will. So let's talk about free will.

Free Will

The problem of free will arises in several contexts. Here's an old one: Since God is omnipotent, it might seem unfair that we be held responsible for *anything*. God, after all, had control. Medieval theologians resolved this issue by deciding that every train of events starts with a "remote efficient cause" and ends with a "final cause," both in God's hands. Causes in between come about through our free choices, for which we will be held accountable on judgment day.

This Medieval concern is not completely remote from that of today's philosophers of morality. Criminal defense lawyers make the concern practical by arguing that the defendant's actions were determined by genetics and environment rather than by free will. We deal with a more straightforward free will issue.

Classical physics, Newtonian physics, is completely deterministic. An "all-seeing eye," knowing the situation of the universe at one time, can know its entire future. If classical physics applied to *everything*, there would be no place for free will.

However, free will can happily coexist with classical physics. In our chapter on the Newtonian worldview, we told how physics, in days gone by, could stop at the boundary of the human body, or certainly at the then completely mysterious brain. Physical scientists could dismiss free will as not their concern and leave it to philosophers and theologians.

That dismissal does not come easily today as scientists study the operation of the brain, its electrochemistry, and its response to stimuli. They deal with the brain as a physical object whose behavior is governed by physical laws. Free will does not fit readily into that picture—it just lurks as a specter off in a corner.

Most neurophysiologists and many psychologists tacitly ignore that corner. Some, perhaps being more logically consistent, deny that free will exists and claim that our *feeling* of free will is an illusion. Others just accept it as a mystery to ignore for now. Yet others explore it. We deal with the controversy this creates when we discuss below the "hard problem" of consciousness.

How could you *demonstrate* the existence of free will? Perhaps all we have is our own feeling of free will and the claim of free will that others make. If no demonstration is at all possible, perhaps the existence of free will is meaningless. (Counter to that argument, though you can't demonstrate your feeling of pain to someone else, you know it exists, and it's certainly not meaningless.)

The most famous experiments bearing on free will have generated fierce argument. In the early 1980s, Benjamin Libet had his subjects flex their wrist at a time of their free choice, but without forethought. He determined the order of three critical times: the time of the "readiness potential," a voltage that can be detected with electrodes on the scalp almost a second before any voluntary action actually occurs; the time of the wrist flexing; and the time the subjects reported that they had made their *decision* to flex (by watching a fast-moving clock).

One might expect the order to be (1) decision, (2) readiness potential, (3) action. In fact, the readiness potential *preceded* the reported decision time. Does this show that some deterministic function in the brain brought about the supposedly free decision? Some, not necessarily Libet, do argue this way. But the times involved are fractions of a second, and the meaning of the reported decision time is hard to evaluate. Moreover, since the wrist action is supposed to be initiated without any "preplanning," the experimental result seems, at best, ambiguous evidence against conscious free will.

Though it is hard to fit free will into a scientific worldview, we cannot ourselves, with any seriousness, doubt it. J. A. Hobson's comment seems apt to us: "Those of us with common sense are amazed at the resistance put up by psychologists, physiologists, and philosophers to the obvious reality of free will."

However, as we have shown, in accepting both free will and the demonstrated quantum results we face an enigma: the apparent creation of reality by conscious observation. Moreover, to avoid the enigma by denying free will, we must also assume that the world conspires to correlate our choices with the physical situations we then observe. The creation of reality by observation is hard to accept, but it is not a new notion.

From Berkeley to Behaviorism

The idea of physical reality being created by its observation goes back thousands of years to Vedic philosophy. But we skip ahead to the eighteenth century. In the wake of Newton's mechanics, the materialist view that all that exists is matter governed by purely mechanical forces gained wide acceptance. Not everybody was happy with it.

The idealist philosopher George Berkeley saw Newtonian thinking as demeaning our status as freely choosing moral beings. That classical physics seemed to leave little room for God appalled him. He was, after all, a bishop. (It was common in those days for English academics to be ordained as Anglican priests, though the celibacy of Newton's day was no longer required—Berkeley married.)

Berkeley totally rejected materialism with the motto *esse est percipi*, "to be is to be perceived," meaning all that exists is created by its observation. To the old question, "If a tree falls in the forest with no one around to hear it fall, is there any sound?" Berkeley's answer would presumably be that there wasn't even a tree were it not observed.

Though Berkeley's almost solipsistic stance may seem a bit batty, many idealist philosophers of the day were enthusiastic about it. Not so for Samuel Johnson, who supposedly responded by kicking a stone, stubbing his toe, and declaring, "I refute him thus!" Stone kicking made little impression on those partial to Berkeley's thinking, which is, of course, impossible to disprove.

Though this is not quite Berkeley's position, here is a centuries-old limerick to illustrate the attention such ideas got:

> There was a young fellow named Todd
> Who said, "It's exceedingly odd
> To think that this tree
> Should continue to be
> When there's no one about in the Quad."

The reply:

> There is nothing especially odd;
> I am always about in the Quad.
> And that's why this tree
> Can continue to be
> When observed by
> Yours faithfully, God.

God may be omnipotent but, we note in the spirit of this limerick, he is not omniscient. If God collapses the wavefunctions of large things to reality by His observation, quantum experiments indicate that He is not observing the small.

The idea that the world around us was being created by its observation never took hold. Most practical people, surely most scientists of the eighteenth century, considered the world to be made up of solid little particles, which some called "atoms." These were presumed to obey mechanical laws much as did those larger particles, the planets. While physical scientists might speculate about the mind, and some used hydraulic pictures for it instead of today's computer models, for the most part they ignored it.

In the nineteenth and much of the twentieth century, scientific thinking was generally equated with materialist thinking. Even in psychology departments, consciousness did not warrant serious study. Behaviorism became the dominant view. People were to be studied as "black boxes" that received stimuli as input and provided behaviors as output. Correlating the behaviors with the stimuli was all that science needed to say about what goes on inside. If you knew the behavior corresponding to every stimulus, you would know all there is to know about the mind.

The behaviorist approach had success in revealing how people respond and, in some sense, why they act as they do. But it did not even *address* the internal state, the feeling of conscious awareness and the making of apparently free choices. According to behaviorism's leading spokesman, B. F. Skinner, the assumption of a conscious free will was unscientific. But with the rise of humanistic psychology in the latter part of the twentieth century, behaviorist ideas seemed sterile.

The "Hard Problem" of Consciousness

Behaviorism had waned when, in the early 1990s, David Chalmers, a young Australian philosopher, shook up brain science by identifying the "hard problem" of consciousness. In a nutshell, the hard problem is that of explaining how the biological brain generates the subjective, inner world of experience. Chalmers's "easy problems" include such things as the reaction to stimuli and the reportability of mental states—and all the rest of consciousness studies. Chalmers does not imply that his easy problems are easy in any absolute sense. They are supposedly easy only relative to the hard problem. Our own interest in the hard problem of consciousness, or awareness, arises from its apparent similarity (and connection?) to the hard problem of quantum mechanics: the problem of observation.

Before going on about the hard problem and the heated arguments it continues to generate, a bit about David Chalmers: As an undergraduate student, he studied physics and mathematics and did graduate work in mathematics before switching to philosophy. Though it is not central to his argument, Chalmers considers quantum mechanics likely to be relevant to consciousness. The last chapter of his landmark book, *The Conscious Mind* (1996), is about the interpretation of quantum mechanics. David Chalmers was a faculty colleague at the University of California, Santa Cruz, in the philosophy department, before he (to our regret) moved to the University of Arizona to become a director of the Center for Consciousness Studies. He is now (at the time of this writing) back in his native Australia and the director of the Centre for Consciousness at Australia National University.

Chalmers's easy problems often involve the correlation of neural activity with aspects of consciousness, the "neural correlates of consciousness." Brain-imaging technology today allows the detailed visualization of activity inside the thinking, feeling brain and has stimulated fascinating studies of thought processes.

Exploration of what goes on inside the brain is not new. Neurosurgeons have long correlated electrical activity and electrical stimulation with reports of conscious perception by placing electrodes directly on the exposed brain. This is done largely for therapeutic purposes, of course, and scientific experimentation is limited. Electroencephalography (EEG), the detection of electrical potentials on the scalp, is even older. EEG can rapidly detect neuronal activity but can't tell where in the brain the activity is taking place.

Positron emission tomography (PET) is better at finding out just where in the brain neurons are firing. Here, radioactive atoms, of oxygen, for example, are injected into the blood stream. Radiation detectors and computer analysis can determine where there is an increase in metabolic activity calling for more oxygen and correlate this with both stimuli and reports of conscious perceptions.

The most spectacular brain imaging technology is functional magnetic resonance imaging (fMRI). It is better than PET at localizing activity and involves no radiation. (The examined head must, however, be held still in a large, usually noisy, magnet.) MRI is the medical imaging technology we described in chapter 8 as one of the practical applications of quantum mechanics. fMRI can almost pinpoint just what part of the brain is using more oxygen during a particular brain function.

fMRI can correlate particular brain regions with the neural processes involved in, say, memory, speech, vision, and reported awareness. The computer-generated, false-color brain images produced can display a bright red spot in particular places in a brain image when someone thinks, say, of food or feels pain. Like any technique based on metabolic activity, fMRI is not fast.

The data today relating neural electrochemistry to consciousness are fragmentary. Just suppose, however, that improved fMRI—or some future technology —could completely identify particular brain activations with certain conscious experiences. This would correlate all (reported) conscious feelings with metabolic activity, and perhaps even with the underlying electrochemical phenomena. Such a complete set of the neural correlates of consciousness is the goal of much of today's consciousness research involving the brain.

Were this goal achieved, we would have accomplished all that *can* be accomplished, some say. Consciousness, they claim, would be explained because there is nothing to it beyond the neural activity we correlate with experienced or reported feelings of consciousness. If we take apart an old pendulum clock and see how the swinging weight driven by a spring moves the gears, we can learn all there is to know about the workings of the clock. Some claim that consciousness will be similarly explained by our learning about the neurons making up the brain.

Francis Crick, physicist co-discoverer of the DNA double helix, who turned brain scientist, has been looking for the "awareness neuron." For him, our subjective experience—our consciousness—is nothing but the activity of such neurons. His book *The Astonishing Hypothesis* (1994) identifies that hypothesis:

> 'You,' your joys and sorrows, your memories and your ambitions, your sense of personal identity and free will, are in fact no more than the behavior of a vast assembly of nerve cells and their associated molecules.

If so, our feeling that consciousness and free will are something beyond the mere functioning of electrons and molecules is an illusion. Consciousness should therefore ultimately have a reductionist explanation, that is, be completely describable in terms of simpler entities, the neural correlates of consciousness. Subjective feelings thus supposedly "emerge" from the electrochemistry of neurons. This is akin to the more readily accepted idea that the wetness of water emerges from the interaction of hydrogen and oxygen atoms forming contiguous molecules of H_2O.

Such emergence forms Crick's "astonishing hypothesis." Is it really so astonishing? We suspect that, to most physicists, at least, it would seem the natural guess.

Crick's long-time younger collaborator, Christof Koch, takes a more nuanced approach:

> Given the centrality of subjective feelings to everyday life, it would require extraordinary factual evidence before concluding that qualia and feelings are illusory. The provisional approach I take is to consider first-person experiences as brute facts of life and seek to explain them.

In a slightly different context Koch further balances different views:

> While I cannot rule out that explaining consciousness may require fundamentally new laws, I currently see no pressing need for such a step.
> ... I assume that the physical basis of consciousness is an emergent property of specific interactions among neurons and their elements. ...
> [But] [t]he characters of brain states and of phenomenal states [of neurons] appear too different to be completely reducible to each other. I suspect that their relationship is more complex than traditionally envisioned.

David Chalmers, the principal spokesperson for a point of view diametrically opposite to Crick's, sees it as impossible to explain consciousness purely in terms of its neural correlates. At best, he maintains, such theories tell us something about the *physical* role consciousness may play, but those physical theories tell us nothing about how it arises:

> For any physical process we specify there will be an unanswered question: Why should this process give rise to [conscious] experience? Given any such process, it is conceptually coherent that it could . . . [exist] in the absence of experience. It follows that no mere account of physical process will tell us why experience arises. The emergence of experience goes beyond what can be derived from physical theory.

While atomic theory might reductively explain the wetness of water and why it clings to your finger, that's a far cry from explaining your *feeling* of wetness. Chalmers, denying the possibility of any reductive explanation of consciousness, suggests that a theory of consciousness should take experience as a primary entity alongside mass, charge, and space-time. He suggests this new fundamental property would entail new fundamental laws, which he calls "psychophysical principles."

Chalmers goes on to speculate on these principles. The one he considers basic, and the one most interesting to us, leads to a "natural hypothesis: that information (at least some information) has two basic aspects, a physical aspect and a phenomenal aspect." This smacks of the situation in quantum mechanics, where the wavefunction also has two aspects. On the one hand, it is the total physical reality of an object, while on the other hand, that reality, some have conjectured, is purely information—whatever that means!

To argue that conscious experience goes beyond ordinary knowing, we are told the story of Mary, a scientist of the future who knows all there is to know about the perception of color. But she has never been outside a room where everything is black or white. One day she is shown something red. For the first time, Mary *experiences* red. Her experience of red is something *beyond* her complete knowledge of red—or is it? You can no doubt generate for yourself the pro and con arguments that the Mary story provokes.

Philosopher Daniel Dennett in his widely quoted book *Consciousness Explained* (1991) describes the brain's dealing with information as a process where "multiple drafts" undergo constant editing, coalescing at times to produce experience. Dennett denies the existence of a "hard problem" as being a form of mind–brain dualism, something he claims to refute by arguing:

> No physical energy or mass is associated with them [the signals from the mind to the brain]. How then do they make a difference to what happens in the brain cells they must affect, if the mind is to have any

influence over the body? . . . This confrontation between quite standard physics and dualism . . . is widely regarded as the inescapable and fatal flaw of dualism.

Since Chalmers argues that consciousness obeys principles *outside* standard physics, it is not clear that an argument based on standard physics can be a refutation of Chalmers. Moreover, there's a quantum loophole in Dennett's argument: No mass or energy is necessarily required to determine to *which* of the several possible states a wavefunction will collapse upon observation.

Our own concern with the hard problem arises, of course, because physics has encountered consciousness in the quantum enigma, which physicists call the "measurement problem"—that is, where aspects of observation come close to those of conscious experience. In each case, something beyond the normal treatment of physics or psychology appears to be needed for a solution. And might those two somethings conceivably be the *same* something?

The essential nature of the observer problem in quantum mechanics has been in dispute since the inception of the theory. Similarly, ever since consciousness has become scientifically discussed in psychology and philosophy, its essential nature has been in dispute. An example of this extreme divergence appeared in early 2005 in the *New York Times,* where some leading scientists were asked to state their beliefs. According to cognitive scientist Donald Hoffman:

> I believe that consciousness and its contents are all that exists. Spacetime, matter and fields never were the fundamental denizens of the universe but have always been, from their beginning, among the humbler contents of consciousness, dependent on it for their very being.

Psychologist Nicholas Humphrey sees it differently:

> I believe that human consciousness is a conjuring trick, designed to fool us into thinking we are in the presence of an inexplicable mystery.

One way to explore the nature of consciousness—and whether it exists—is to ask who or what can possess it.

A Conscious Computer?

We each *know* we are conscious. Perhaps the only evidence for believing that others are conscious is that they more or less look like us and behave like us. What other evidence is there? The assumption that our fellow humans are conscious is so ingrained that it is hard to express the reasons for our believing it.

How far down does consciousness extend? What about cats and dogs? What about earthworms or bacteria? Some philosophers see a continuum and even attribute a bit of consciousness to a thermostat. On the other hand, maybe consciousness turns on abruptly at some point on this scale. After all, Nature can be discontinuous—going below 32° F, liquid water abruptly becomes solid ice.

Let's step back from consciousness and just talk about "thinking," or intelligence. Today, computer systems called artificial intelligence, or AI, assist doctors in diagnosing disease, generals in planning battles, and engineers in designing yet better computers. In 1997, IBM's Deep Blue beat the world chess champion, Garry Kasparov.

Did Deep Blue *think?* It depends on what you mean by thinking. Information theorist Claude Shannon, when asked whether computers will ever think, supposedly replied: "Of course. I'm a computer, and I think." But the IBM scientists who designed Deep Blue insist that their machine is just a fast calculator evaluating a hundred million chess positions in the blink of an eye. Whether or not it thinks, Deep Blue is surely not conscious.

But if a computer *appeared* conscious in every respect, wouldn't we have to accept it as conscious? We should follow the time-honored principle that if it looks like a duck, walks like a duck, and quacks like a duck, it must be a duck.

The interesting question is whether it is possible to *build* a conscious computer, therefore a conscious robot. Computer consciousness is sometimes called "*strong* AI." (Would it be murder to pull the plug on a truly conscious robot?) Logical "proofs" have been advanced that strong AI is, in principle, possible. There are other "proofs" that it is impossible. How could you tell if a computer were conscious?

In 1950 Alan Turing proposed a test for computer consciousness. He actually called it a test for whether a computer could think; a scientist wouldn't use the term "consciousness" back then. (Turing also designed the first programmed computer and developed a theorem for what computers could ultimately do, or not do. Turing was later arrested for his homosexuality, and in 1954 committed suicide. Many years after his death, officials revealed that it was Turing who broke Germany's Enigma code. The Allies were thus able to read the enemy's most secret messages, probably shortening World War II by many months.)

The Turing test uses essentially the same criterion for deciding whether a computer is conscious as we do in ascribing consciousness to another individual: Does it look and behave more or less like me? Let's not worry about the "look" part; a human-*looking* robot can no doubt be accomplished. The issue is whether its computer brain gives it consciousness.

To test whether a particular computer is conscious, it should, according to Turing, be enough to communicate with it by a keyboard and carry on any conversation, for as long as you wish. If you can't tell whether you are communicating with a computer or another human, it passes the Turing test. Some would then say that you cannot deny that it is conscious.

In class one day, I (Bruce) casually commented that any human could easily pass a Turing test. One young woman objected: "I've *dated* guys who couldn't pass a Turing test!"

Consciousness is a mystery we explore because physics' encounter with it presents us with the quantum enigma. In our next chapter, the mystery meets the enigma.

16

The Mystery Meets the Enigma

*When the province of physical theory was extended
to encompass microscopic phenomena through the
creation of quantum mechanics, the concept of
consciousness came to the fore again: It was not
possible to formulate the laws of quantum mechanics
in a fully consistent way without reference to the
consciousness.*

—Eugene Wigner

*When there are two mysteries, it is tempting to
suppose that they have a common source. This
temptation is magnified by the fact that the problems
in quantum mechanics seem to be deeply tied to the
notion of observership, crucially involving the relation
between a subject's experience and the rest of the
world.*

—David Chalmers

Consciousness and the quantum enigma are not just two mysteries; they are *the*
two mysteries: the first, our physical demonstration of the quantum enigma, faces
us with a fundamental mystery of the objective world "out there;" the second,
conscious awareness, faces us with the fundamental mystery of the subjective,
mental world "in here." Quantum mechanics appears to connect the two.

The Encounter "Officially" Proclaimed

In his rigorous 1932 treatment *The Mathematical Foundations of Quantum Mechanics,* John von Neumann showed that quantum theory makes physics' encounter with consciousness inevitable. He considered a measuring apparatus, a Geiger counter, for example. It is isolated from the rest of the world but makes contact with a quantum system, say, an atom simultaneously in two boxes. This Geiger counter is set to fire if the atom is in the top box and to remain unfired if the atom is in the bottom box. Von Neumann showed that since the Geiger counter is a physical system governed by quantum mechanics, it would enter the superposition state with the atom and be, simultaneously, in the fired and unfired state. (We saw this situation in the case of Schrödinger's cat.)

Should a second isolated measuring apparatus come into contact with the Geiger counter—for example, an electronic device to record whether the Geiger counter has fired—it joins the superposition state and records both situations as existing simultaneously. This so-called "von Neumann chain" can continue indefinitely. Von Neumann showed that *no* physical system obeying the laws of physics (i.e., quantum theory) could collapse a superpostion state wavefunction to yield a particular result. (Nonetheless, for all practical purposes one can, of course, *consider* the wavefunction collapsed at any macroscopic stage of the von Neumann chain.)

However, when we look at the Geiger counter, we will always see a particular result, not a superposition. Von Neumann concluded that only a conscious observer doing something that is not encompassed by physics can collapse a wavefunction. Only a conscious observer can actually make an observation.

A couple of years later Schrödinger told his cat story to illustrate the "absurdity" of his own quantum theory. His cat story was essentially based on von Neumann's conclusion requiring someone to consciously observe in order to collapse a superposition state. In this sense, physics, the most basic empirical science, is based on consciousness.

Consciousness and Reduction

With the reductionist perspective, one seeks to reduce a complex system to underlying science. For example, one seeks explanations of psychological phenomena in biological terms. Biological phenomena can then be seen as ultimately chemical. And no chemist doubts that chemical phenomena are fundamentally the interactions of atoms obeying quantum physics. Physics, itself, supposedly rests firmly on primitive empirical ground.

In chapter 4 we represented this view with the reductionist pyramid. We now see that the classical view of the primitive empirical ground on which physics rests is challenged by quantum mechanics. In some basic sense, physics rests on the phenomenon of wave-function collapse by conscious observation. Therefore, we add a somewhat cloudy consciousness at the base of our reductionist pyramid. While, for all practical purposes, science will always be hierarchical, with each level in the hierarchy needing its own set of concepts, the new view of reduction may change the way we perceive the whole scientific enterprise.

But we won't treat this epistemological issue for now.

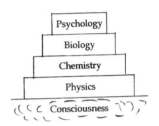

Figure 16.1 Hierarchy of scientific explanation revisited

Conscious Awareness versus Entanglement

Consider once more our atoms in a box pair. We will come to a strange question about consciousness. As we said in chapter 14, we can demonstrate that a photon coming through one of the boxes does not observe whether or not the atom is in that box. It rather joins a superposition state with the atom. Should the photon hit an *isolated* Geiger counter, that counter just joins the atom–photon superposition state (according to von Neumann's analysis). It is simultaneously fired and unfired, and the atom is still simultaneously in both boxes.

Let's consider a different situation. Suppose that the counter was sitting on a table that rests on the floor. This *not*-isolated counter is thus entangled with the table, and therefore with the rest of the world—which includes people. The atom, entangled with the photon, which entangled with the counter, is now entangled with conscious individuals. Nevertheless, if no one looks, no one *knows* which box the atom is in.

Here's our promised strange question: Does the mere entanglement with conscious individuals collapse the atom wholly into a single box? Or does the collapse into a single box require conscious awareness of which box the atom is in by an actual look at the atom or the Geiger counter? How could we possibly tell whether the atom has collapsed into a single box or is still simultaneously in both? Strictly speaking, without looking at the atom or the Geiger counter, we can't. So, although it might seem silly, the atom is perhaps still in both boxes.

We can say a bit more about awareness and entanglement. Everything in the

Figure 16.2 Drawing by Nick Kim, 2000. © American Institute of Physics

world immediately entangles with our photon as soon as it hits the not-isolated Geiger counter. Entanglement travels infinitely fast. But for a distant person to become aware of the condition of the counter or the atom, he or she would have to communicate by some physical means that could not exceed the speed of light. Awareness can travel no faster than the speed of light.

We saw entanglement travel faster than light, infinitely fast, presumably, in Bell's theorem experiments. Immediately upon the observation of the polarization of one twin-state photon, the polarization of its twin is set—that's entanglement. But only when the two observers gain awareness of each other's result can they know whether they had a match or mismatch. A photon "learns" its twin's behavior instantaneously, but Alice and Bob (two conscious observers) can become aware of each other's result only at a rate limited by the speed of light.

Here's a cartoon strip from *Physics Today* in May 2000 that's relevant in a few ways. (When the quantum enigma comes up in physics journals, other than to supposedly resolve the issue, it's often treated with humor.)

Chris, being entangled with Eric and the rest of the world, would, of course, not go into a "superposition of all possible states" when Eric looks away. After all, an atom you found in a particular box would not go into a superposition in both boxes when you looked away. Moreover, assuming Chris is a conscious observer, not an isolated robot, she would constantly have awareness of her own body's position and constantly collapse her wavefunction.

Do We Need a *Conscious* Observer?

A robot argument is often presented to evade physics' encounter with consciousness. Here's how it goes: We don't need a conscious observer to collapse a wavefunction, because a not-conscious, isolated robot can do it. For example, a robot could be presented with sets of our box pairs. The robot could randomly choose to do either a look-in-a-box experiment or an interference experiment and print out a report of its results. Since that printout would be indistinguishable from an observation presented by a conscious observer, the not-conscious robot can therefore be considered to be an observer. No consciousness need be involved.

Does this argument work? Consider whether this robot-performed experiment avoids the encounter with consciousness from a human perspective, the only meaningful perspective. You're given the robot's printout. The printout indicates, for example, that the robot did a look-in-the-box experiment with box-pair sets 2, 5, 7, 8, 11, and 13, demonstrating that these sets contained objects wholly in a single box. And the printout shows that for sets 1, 3, 4, 6, 9, 10, and 12 it did an interference experiment, demonstrating that these sets contained objects distributed over both boxes. In itself, the robot's printout presents no problem, displays no enigma.

In fact, receiving the robot's printout, you might well assume that the box-pair sets indeed contained objects of just such kinds. You could assume that in the necessarily unobserved preparation process the objects were *created* either distributed or concentrated.

However, you wonder, if the box-pair sets were indeed different in that way, how did the robot "decide" to do the appropriate experiment with each box-pair set? If it did an interference experiment with objects wholly in a single box, it would get no pattern, just a uniform distribution of objects. And what would it see and report if it did a look-in-a-box experiment with objects physically distributed over both boxes? A partial object is never reported. The robot's supposedly random choice was strangely always the "right" choice.

You're prompted to investigate how the robot chose which experiment to do in each case. Suppose you find that it flipped a coin. Heads, it did the look-in-the-box experiment; tails, the interference experiment. You find something puzzling about this: The coin's landing seems inexplicably connected with what was presumably in a particular box-pair set. Unless ours is a strangely deterministic world, one that conspired to correlate the coin landing with what was in the box pairs, there is no physical mechanism for that correlation.

You therefore replace the robot's coin flipping by the one decision mechanism that you are sure is not connected with what supposedly exists in a particular box-pair set: *your own free choice*. You push a button telling the robot which experiment to do with each box-pair set. You now find that by your conscious free choice of experiment you can prove either that the objects were concentrated or that they were distributed. You can choose to prove either of two contradictory things. You are faced with the quantum enigma, and consciousness is involved.

Our robot discussion did not involve quantum theory. We intended it to be theory-neutral, as was our description of the box-pairs experiment itself. If we consider the robot argument from the view of quantum theory, the isolated robot is a quantum system, and von Neumann's conclusion applies: The robot entangles with the object in the box pairs; and the object's wavefunction does not collapse into a single box until a conscious observer views the robot's printout.

The Only *Objective* Evidence for Consciousness

By "objective evidence" we mean third-person evidence that can be displayed to essentially anybody. Objective evidence in this sense is the normal requirement for establishing the reliability of a scientific theory. We each know we are conscious; that's first-person evidence for consciousness. Others report they are conscious; that's second-person evidence. Without third-person evidence, objective evidence that consciousness itself can do something physically observable, its existence is deniable. As we saw in chapter 15, some claim that "consciousness" is no more than a name for the electrochemical behavior of the vast assembly of nerve cells and their associated molecules in our brains.

What might qualify as objective evidence that consciousness is more than that, that it can do something physical? Does the box-pair experiment qualify? The evidence provided by the quantum experiment is circumstantial. That is, one fact (interference) is used to establish a second fact (the object in both boxes). Circumstantial evidence can be convincing beyond a reasonable doubt. (It can, for example, legally secure a conviction.) But its logic can be circuitous. We therefore present a parable displaying *direct* evidence for consciousness.

Our parable tells of something much like what our visitor saw in Neg Ahne Poc. It is an impossible story, but its direct evidence is easy to analyze, and the analogy with our quantum experiment puts that circumstantial evidence more clearly in front of us. (We originally included this story in an article, titled "The Only Objective Evidence for Consciousness," that we wrote for the *Journal of Mind and Behavior*.)

Our imaginary Dr. Elbe claims to demonstrate psychokinesis, or PK. (PK is physical phenomena external to the body brought about by conscious mental effort alone, that is, without any physical interaction.) Dr. Elbe displays a large number of box pairs. She instructs you, in her first experiment, to determine which box of each pair holds a marble by opening the boxes of a pair in turn. Opening the boxes sequentially, about half the time you find a marble in the first box you open and half the time in the second. You conclude that immediately prior to your observation one box of each pair contained a marble.

Noting that each marble can come apart into white and black hemispheres, Dr. Elbe presents a second set of box pairs. She now instructs you to determine which box of each pair contains the white hemisphere and which the black by opening both boxes of each pair *at about the same time.* Opening the boxes simultaneously, you always find a white hemisphere in one of the boxes and a black in the other box of that pair. You conclude that for *this* set of box pairs, immediately prior to your observation, a marble had been distributed over both boxes of each pair.

Now presenting you with further sets of box pairs, Dr. Elbe suggests that for each set *you* choose either of the two previous experiments, that is, you choose whether to open the boxes sequentially or simultaneously. Allowed to repeat the experiment *of your choice* as many times as you wish, you *always* observe a physical result correlated with the experiment you choose. Whenever you decide to open the boxes sequentially, you find the marble wholly in a single box of a pair; whenever you decide to open the boxes simultaneously, you find the marble distributed over both boxes of the pair.

Puzzled, you challenge Dr. Elbe: "Obviously, some of your sets of box pairs had a whole marble in a single box, while other box-pair sets contained half of a marble in each box. But how did we always get a result corresponding to the opening method I chose? What if I had made the other choice with a particular set of box pairs? After all, before I opened the boxes, each marble had to be either wholly in a single box or distributed over both boxes of the pair. When you gave me a set of box pairs, how did you know which experiment I would choose?"

Dr. Elbe responds: "I did *not* know which experiment you would choose. Your conscious choice *created* the particular situation of the marble in its box pair. The condition of the marble would have been different had you made a different choice. You have seen consciousness displayed as a physically efficacious entity beyond its neural correlates—what we call PK."

You are sure there's trickery involved. After all, Dr. Elbe's demonstration involved more than an expression of conscious intent. It required you to move your hands and open boxes. Perhaps the mechanical opening of the box pairs, either sequentially or simultaneously, somehow physically put the marble wholly in a single box or spread it over two boxes.

Therefore, with your unlimited resources, you bring in a broad-based team of scientists and magicians (illusionists) to investigate Dr. Elbe's demonstration—with particular emphasis on the hand motion and the box-opening technique. However, after a study you accept as exhaustive, they report there to be no trickery and that *no physical explanation* could be found.

Of course, Dr. Elbe's demonstration cannot be done. But *if* it could, you would be compelled to accept it, at least, as objective evidence that conscious choice itself can affect a physical situation, evidence that consciousness exists as an entity beyond its neural correlates.

The archetypal quantum experiment, the two-slit experiment or our box pairs experiment, comes very, very close to Dr. Elbe's demonstration. Your conscious decision of which experiment to do (look in the box or interference) created either of two contradictory prior physical situations in the box pairs. The quantum experiment is thus objective evidence for consciousness. Evidence, of course, is not proof. But the quantum experiment is the *only* objective evidence for consciousness. We thus point to a footprint at the crime scene without naming a culprit.

Nevertheless, it is with trepidation that we *speculate* that such a quantum experiment shows consciousness reaching out and doing something physical. In serious moments as physicists, we can't even half-believe that. But we're hardly the first to so speculate. Developer of quantum theory and Nobel laureate Eugene Wigner has written:

> Support [for] the existence of an influence of the consciousness on the physical world is based on the observation that we do not know of any phenomenon in which one subject is influenced by another without exerting an influence thereupon. This appears convincing to this writer. It is true that under the usual conditions of experimental physics or biology, the influence of any consciousness is certainly very small. "We do not need the assumption that there is any such effect." It is good to recall, however, that the same may be said of the relation of light to mechanical objects. . . . It is unlikely that the [small] effect would have been detected had theoretical considerations not suggested its existence. . . .

Position Is Special

Why can't we *see* an object simultaneously in two boxes? Quantum theory provides no answer. Strictly speaking, an object wholly in Box A can also be considered to be in a "superposition state." It is in a superposition (or sum) of the state {in Box A + in Box B} plus the state {in Box A − in Box B}. Notice that these just add up to {in Box A}. Similarly, the living-cat state is a superposition of the state {living + dead} plus the state {living − dead}. The missing factor of 2 is accounted for in the actual mathematics of quantum theory.

All these states have equivalent status as far as quantum theory is concerned. Why, then, do we always see things in certain kinds of states—states characteristic of a particular position? We never actually see the weird states corresponding to things being simultaneously in different positions. (Schrödinger's simultaneously living-and-dead cat is such a weird state because some of the atoms in a living cat must be in different positions than the atoms in a dead cat.)

For our object in a box pair we inferred that it had been in a superposition state, simultaneously in different boxes, by doing an interference experiment. But our *actual* experiences in the interference experiment were the positions of objects in particular interference maxima.

Arguably, the reason we observe only states characterized by unique positions is that we humans are beings who can experience *only* position (and time). Speed, for example, is position at two different times. When we see things with our eyes, it is because of light on particular positions on our retina. We feel by touch the position of something on our skin; we hear by the changing position of our eardrums; we smell by the effects on certain receptor positions in our nose. We therefore build our measuring instruments to display their results in terms of position—typically, that of a meter pointer or a light pattern on a screen. Nothing in quantum theory forces this situation. We humans seem constructed in this special way.

Is it conceivable that other beings could experience reality differently? Could they possibly directly experience the superposition states whose existence we can only infer? To them, an atom simultaneously in both boxes, or Schrödinger's cat simultaneously alive and dead, would be "natural." That is, after all, the quantum way, presumably Nature's way. They would therefore experience no measurement problem, no quantum enigma.

Two Enigmas

There are actually two measurement problems, two enigmas. We focused on observer-created reality, observation causing, say, the looked-at atom to appear wholly in a single box, or Schrödinger's cat to be either alive or dead. (Einstein's quip that he believed the moon was really there even when no one was looking referred to this enigma.) The less-serious enigma is Nature's randomness: How does it come about that the atom randomly appears in, say, Box A rather than Box B? How does the cat randomly come to be in, say, the alive state? (Einstein's quip that God doesn't play dice referred to this enigma.)

With Everett's many-worlds interpretation, we choose all possible experiments and see all possible results. According to this view, we are troubled by two enigmas only because we do not realize that at every observation we split and simultaneously exist in a multitude of different worlds. From an Everettian point of view, we see neither enigma.

Let's contrast the two enigmas with a bit of fantasy (inspired by a parable by Roland Omnès). On the higher plane on which they dwell, Everettians happily experience the multitude of simultaneous realities given by quantum theory. No enigmas trouble them. One young Everettian, sent down to explore planet Earth, was shocked to find his simultaneous multiple realities collapse to a single actuality. His curiosity impelled repeated descents. Each time he saw his realities randomly collapse to one of the many actualities he was accustomed to perceive simultaneously on his higher plane. Baffled by this collapse, something not explicable within the quantum theory he understood so well, he reported an enigma: Down on Earth, Nature randomly selects a single actuality.

Our Everettian had a favorite way of looking at the multiple realities he could experience. He understood, however, that this personal choice, or "basis," was logically equivalent to any other. In a rather unusual mood on a particular descent to Earth, our Everettian adopted a different basis for his multiple realities. He experienced a second shock. The collapse was not only to a specific actuality, something he had now gotten accustomed to, but to an actuality that was logically inconsistent with one presented by his previous way of looking. He had to report a second, and more troubling, enigma: Down on Earth, his conscious choice of the way he looked created *inconsistent* realities.

Analogies

Whether or not a consciousness itself can have physical impact beyond its own brain, there are compelling analogies between quantum mechanics and consciousness. Analogies, of course, prove nothing, but they can stimulate and direct thinking. Analogies with Newton's mechanics sparked the Enlightenment. Here's a very general one by Niels Bohr:

> [T]he apparent contrast between the continuous onward flow of associative thinking and the preservation of the unity of the personality exhibits a suggestive analogy with the relation between the wave description of the motions of material particles, governed by the superposition principle, and their indestructible individuality.

Here are a few more:

Duality: It is often argued that the existence of consciousness cannot be deduced from properties of the material brain. Two qualitatively different processes seem to be involved. In quantum theory, an actual event comes about not by the evolving wavefunction, but by the collapse of the wavefunction by observation. Two qualitatively different processes seem involved.

"Nonphysical" influences: If there's a mind that's other than the physical brain, how does it communicate with the brain? This mystery recalls the connection of two quantum-entangled objects with each other—by what Einstein called "spooky actions" and Bohr called "influences."

Observer-created reality: Berkeley's "to be is to be perceived" is the preposterous solipsistic view of the effect of consciousness. But it is reminiscent of what happens with our object in a box pair or with Schrödinger's cat.

Observing one's thoughts: If you think about the content of a thought (its position), you inevitably change where it is going (its motion). On the other hand, if you think about where it is going, you lose the sharpness of its content. Analogously, the uncertainty principle shows that if you observe the position of an object, you change its motion. On the other hand, if you observe its motion, you lose the sharpness of its position.

Parallel processing: Neuronal action rates are billions of times slower than those of computers. Nevertheless, with complex problems, such as chess, for example, human brains can still compete with the best computers. The brain presumably achieves its power by working on many paths simultaneously. Such massively parallel processing is what computer scientists attempt to achieve with quantum computers, whose elements are simultaneously in superpositions of many states.

The analogies between consciousness and quantum mechanics lead one to expect that an advance in the fundamentals of one field might stimulate an advance in the other. Analogies might even suggest testable connections of the two.

Two Quantum Theories of Consciousness

Theories encompassing mind and matter that go beyond analogy must be big and bold and are invariably controversial. The Penrose-Hameroff approach is based on quantum gravity, which is the theory needed to describe black holes and the Big Bang, to which Roger Penrose is a major contributor. The Penrose-Hameroff approach to consciousness also involves ideas from mathematical logic and neuronal biology.

The mathematician Kurt Gödel proved that any logical system contains propositions whose truth cannot be proven. We can, however, by insight and intuition, know the answer. Penrose controversially deduces from this that conscious processes are noncomputable. That is, a computer cannot duplicate them. Penrose thus denies the possibility of strong AI. If so, consciousness, like the quantum enigma, goes beyond anything our present science treats.

Penrose proposes a physical process beyond present quantum theory that rapidly collapses macroscopic superpositions to actualities. It causes the object simultaneously in both Box A *and* Box B to become either in Box A *or* Box B. It causes Schrödinger's cat simultaneously both alive *and* dead to become either alive *or* dead. In general, it causes "and" to become "or." This process collapses, or "reduces," the wavefunction objectively, for everybody, even without an observer. Penrose calls this process "objective reduction," abbreviated OR. He notes the appropriateness of the OR acronym—it brings about the "or" situation.

Penrose speculates that OR occurs spontaneously whenever two space-time geometries, and therefore gravitational effects, differ significantly. Stuart Hameroff, an anesthesiologist, who points out that he regularly turns consciousness off and then back on, suggested how this process may occur in the brain. Two states of certain tubulins (proteins) that exist within neurons might display Penrose's OR on a time scale appropriate for neural functions. Penrose and Hameroff claim that superposition states and long-range coherence might exist within a brain even though it is in physical contact with the environment, and that spontaneous ORs could regulate synaptic/neural functions.

Such ORs would constitute "occasions of experience." If entangled with an object external to the observer, the OR would collapse the wavefunction of the observed object, and everything entangled with it.

The three bases of the Penrose-Hameroff theory—noncomputability, the involvement of quantum gravity, and the role of tubulins—are each controversial. And the entire theory has been derided as having the explanatory power of "pixie dust in the synapses." However, unlike most theories of consciousness, quantum or otherwise, it proposes a specific physical mechanism, some fundamental aspects of which seem almost testable with today's technology.

With another theory, Henry Stapp argues that classical physics can never explain consciousness, but an explanation comes about naturally with quantum mechanics. We saw earlier how free will was permitted in deterministic classical physics only by excluding the mind from the realm of physics. Stapp notes that extending such classical physics to the brain/mind would have our thoughts controlled "bottom-up" by the deterministic motion of particles and fields. There would be no mechanism for a "top-down" conscious influence.

Stapp takes off from von Neumann's formulation of the Copenhagen interpretation. Von Neumann, recall, showed that in viewing a microscopic object in a superposition state, the entire measurement system—from, say, the Geiger counter, to the human eye looking at it, to the thus entangled synapses in the observer's brain—must, strictly speaking, be considered part of a grand superposition state. Only a consciousness, something beyond Schrödinger equation evolution, can collapse a wavefunction.

Stapp postulates two realities, a physical and a mental. The physical includes the brain, perhaps in a particular superposition state. The mental reality includes consciousness, thoughts, and, in particular, intentions. The mental can intentionally act on the physical brain to choose a particular superposition state, which then collapses to a particular situation. Consciousness does not itself "reach out" to the external world in this theory, but this mental choice itself nevertheless determines, in part, the character of the physical world external to the body. The final random aspect of the choice (which particular box the object is in, for example) is then made by Nature.

How can a large, warm brain remain in a particular quantum state long enough for a person's intentions to influence it? Stapp answers this with the "quantum Zeno effect." (Named for a Zeno-like claim: A watched pot never boils.) When a quantum system decays from an upper state to a lower, the decay starts very slowly. If it is observed very soon after the decay has started, it will almost certainly be found in the original state. The decay then starts over again from the original state. If the system is observed almost constantly, it almost never decays. Stapp applies this to the mental intentions "observing" the brain and thus holding it in a given quantum state for a sufficient time.

Stapp cites various psychological findings as evidence for his theory. The theory is, of course, controversial.

The Psychological Interpretation of Quantum Mechanics

Though quantum theory is outrageously counterintuitive, it works perfectly. Since Nature need not behave in accord with our intuition, is the measurement problem, the quantum enigma, just in our heads? Maybe. But, if so, why do we find quantum mechanics so hard to accept? Why do the observed facts produce a cognitive dissonance?

Merely to say that we evolved in a world where classical physics is a good approximation is not enough. We evolved in a world where the sun apparently moved across the sky and Earth stood still. Nevertheless, the once-counterintuitive Copernican picture is readily accepted despite our evolution. We also evolved in a world where things moved slowly compared to the speed of light. Einstein's relativity can at first be grossly counterintuitive. Though it is difficult for physics students initially to accept that time passes more slowly in a moving rocketship, they soon do so. We find no "interpretations" of relativity. The more deeply you think of relativity, the less strange it seems. The more deeply you think of quantum mechanics, the more strange it seems.

What is it about the organization of our brain that makes quantum mechanics seem so weird? With this question, most physicists would assign the quantum enigma to psychology. Its resolution would display our unease with physical reality being created by its observation as merely a psychological hang-up. That would be the psychological interpretation of quantum mechanics.

Paraphenomena

Paraphenomena are (supposed) happenings that are presumed inexplicable within normal science. Here are three examples involving the mind: (1) Extra-sensory perception (ESP) is the acquisition of information by some means other than the normal senses, for example, mental telepathy or remote viewing. (2) Pre-cognition is being able to know what will happen in the future. (3) Psychokinesis (PK) is the causing of a physical effect by mental action alone, for example, Uri Geller's spoon bending or the mental influencing of radioactive decay.

According to polls, well more than half of Americans (and English) have significant beliefs in the reality of such phenomena. When asked with a positive spin, "Who thinks that at least a *little* bit of ESP likely exists?" well more than half of the students in a large general physics class raised their hands. (The two of us would answer "*not* likely.")

That widespread acceptance of paraphenomena is sufficient reason for including some comment in our book. A more important reason is that certain competent researchers claiming to display such phenomena cannot be dismissed out of hand. But hard-to-believe things require strong evidence. If someone tells you that there is a black dog outside, you likely just accept it. If they tell you there is a green giraffe, you want to go see for yourself. As yet, evidence for the existence of paraphenomena strong enough to convince skeptics does not exist.

But if—*if!*—such a phenomenon were convincingly demonstrated, we would know where to start looking for an explanation: the quantum effects of consciousness, Einstein's "spooky interactions."

In the following chapter we consider the implications of the quantum enigma on the grandest scale of all, the entire universe.

17

Consciousness and the Quantum Cosmos

In the beginning there were only probabilities. The universe could only come into existence if someone observed it. It does not matter that the observers turned up several billion years later. The universe exists because we are aware of it.

—Martin Rees

We wonder how literally Martin Rees, Cambridge University professor and England's Astronomer Royal, means what he said in this opening quote. Having come this far in the book, you at least know what stimulates his comment. Though quantum mechanics supposedly applies to everything, it's a big step from those things for which observer-created reality has been demonstrated to the whole universe.

Einstein's theory of gravity, "general relativity," appears to work perfectly for the large-scale universe; it tells of black holes and is needed in dealing with the Big Bang. Understanding black holes and the Big Bang also requires understanding things at the small scale. It therefore requires quantum mechanics. Requiring *both* general relativity *and* quantum mechanics poses a problem: General relativity resists connection with quantum mechanics. String theorists and others have tried for decades to couple these two fundamental descriptions of Nature to produce a quantum theory of gravity.

When, some years ago, I told a string-theorist colleague of my interest in the quantum enigma, "Bruce, we're not ready for that," was his response. His point was that progress in what he would call the quantum measurement problem likely required still-to-come advances in quantum gravity theory. Maybe. But, while today's cosmology brings the quantum enigma ever more to the fore, it presents the same enigma, only on an ever-grander scale. In this chapter we look at this big picture to see how the creation of reality by conscious observation has been applied to the universe as a whole.

Black Holes, Dark Energy, and the Big Bang

Black Holes

When a star exhausts the nuclear fuel that keeps it hot and expanded, it collapses under its self-gravitational attraction. If its mass exceeds a critical amount, no force can halt the continuing collapse. General relativity predicts its collapse to a massive, infinitesimal point, a "singularity." Physicists shun singularities, and quantum theory would replace the singularity with an extremely compact, but finite-sized, mass in some not-yet-understood way.

At a distance from this compact mass, within the so-called "horizon," the gravitational attraction is so great that not even light can escape. This collapsed star thus emits no light—therefore, it's black. Anything venturing inside the horizon can never get out—it's a black hole.

Stephen Hawking showed that quantum mechanics enters the black-hole picture not only at the singularity but also at the horizon: Quantum effects should cause the black-hole horizon to emit what is now called "Hawking radiation." Therefore, any black hole that does not suck in mass from its surroundings should eventually radiate away, or "evaporate."

Though the evaporation time scale would be longer than the age of the universe, it has raised a paradox: Quantum theory insists that "information" is always preserved, but if Hawking radiation were random, as initially thought, the information contained in an object falling into a black hole would be lost when the black hole evaporated.

We are using a far-fetched notion of information here. If, for example, you throw your diary into the fire, someone can, in *principle,* recover its information by analyzing the smoke and ashes. The apparent information loss in black-hole evaporation led Hawking to speculate that the information might, when the hole evaporated, be channeled to a parallel universe. (This reminds us of the many-worlds interpretation and provides fodder for science fiction writers.)

Hawking recently decided that the black hole's radiation is *not* random, that the radiation carries off the information dropped into the hole—like smoke carries off information from your burning diary. No need for parallel universes to take up black-hole information. Nevertheless, some cosmologists, for other reasons, suggest that ours is likely not the only universe.

Dark Energy

Modern cosmology is based on Einstein's theory of general relativity. It is "general" in the sense that it extends his earlier special relativity to include accelerated motion and gravity with the realization that the two are equivalent. For example, when the elevator cable breaks, your downward acceleration would cancel your experience of gravity.

Though mathematically complex, general relativity is a conceptually straightforward and beautiful theory. But in the form Einstein first wrote down in 1916, it seemed to have a serious problem. It said the universe could not be stable. The mutual gravitational attraction of the galaxies would cause them to collapse in on themselves. Einstein patched up his theory by adding the "cosmological constant," a repulsive force to counter the gravitational attraction.

In 1929 Astronomer Edwin Hubble announced that the universe was *not* stable—it was, in fact, expanding. The more distant a galaxy, the faster it moved away. If so, some time in the past everything was clumped together, giving rise to the idea of the universe starting with a big explosion, the Big Bang. That presumably explained why galaxies did not fall in on each other. No repulsive force—no cosmological constant—was needed.

An explosion is not quite the right picture. General relativity has space *itself* expanding, not galaxies flying apart in a fixed space. Specks of paper pasted on an inflating balloon, and thus moving apart faster the more distant they are from each other, is a good analogy.

When Einstein realized that the universe was indeed *not* stable, he threw out his cosmological constant, calling it the "greatest blunder of my career." He realized that if he had only believed his original beautiful theory, he could have predicted an expanding (or contracting) universe more than a decade before its discovery.

The gravitational attraction of the galaxies for each other should slow the expansion, just as gravity slows an upward-thrown stone as it rises. The stone rises to some height and then falls back down. Similarly, one might expect the galaxies to reach some maximum separation and eventually fall back together in the Big Crunch.

If you throw a stone up *fast* enough, it will continue out in space forever. However, still pulled back by Earth's gravitational attraction, it will continually slow. By the same token, if the Big Bang were violent enough, the universe would expand forever, albeit at a slower and slower rate. By determining the rate at which an upward-thrown stone is slowing, you can tell whether it will fall back down or continue out forever. By finding the rate at which the expansion of the universe slows, we can tell whether or not to expect the Big Crunch.

Actually, it has been recognized for a couple of decades that the galaxies do not constitute all the mass of the universe, not even the largest part. The motions of stars within galaxies, and other evidence, tell us that there is a kind of matter out there in addition to the stuff that the stars, the planets, and we are made of. It has gravitational attraction but does not emit, absorb, or reflect light. We thus cannot see it—it's "dark matter." No one knows what it is, but people have built detectors to search for the likely suspects. It is the sum of the normal matter and the dark matter that would be expected to slow the expansion and determine the eventual fate of the universe.

(On a recent PBS *Nova* program, an astronomer said he could not think of a more fundamental question for humankind than, "What is the end of the universe?" Perhaps that *is* a pressing question. But it recalls a story: In a popular lecture, an astronomer concluded: "Therefore, in about five billion years the sun will expand as a red giant and incinerate the inner planets, including Earth." "Oh, no!" moaned a man in the rear. "But, sir, it won't happen for another *five billion years*," reassured the astronomer. The man's relieved response was, "Oh, thank God! I thought you said five *million* years.")

In the past decade, astronomers set out to determine the fate of the universe by measuring how fast certain distant exploding stars—supernovas—are receding. These particular explosions have a characteristic intrinsic brightness, and therefore astronomers can tell how far away they are by how bright they appear. And the farther away they are, the longer ago the light we now receive must have left them. Putting all this together, they could determine how fast the universe was expanding at different times in the past, and therefore determine the rate of slowing.

Surprise! The expansion of the universe is *not* slowing—it's accelerating. Not only is the mutual gravitational attraction of the galaxies canceled, but there is a repelling force in space that is *greater* than the gravitational attraction. With that force must come an energy.

Since mass and energy are equivalent ($E = mc^2$), this mysterious repulsive energy has a mass distributed in space. In fact, *most* of the universe is made up of this mysterious "dark energy." The universe appears to be about 70% dark energy and 25% dark matter. The kind of stuff we, the planets, and the stars are made of appears to be a mere 5% of the universe.

Though no one knows what the dark energy is, in a formal sense it brings Einstein's cosmological constant, his "biggest blunder," back into the equations of general relativity. Theoretical guesses have an uncanny way of ending up right.

Is it conceivable that the mysterious dark energy involves the connection between the large-scale universe and consciousness that Rees's comment at the start of this chapter might imply? Let us quote the quantum theorist Freeman Dyson, writing before the idea of dark energy arose:

It would not be surprising if it should turn out that the origin and destiny of the energy in the universe cannot be completely understood in isolation from the phenomena of life and consciousness. . . . It is conceivable . . . that life may have a larger role to play than we have imagined. Life may have succeeded against all odds in molding the universe to its purposes. And the design of the inanimate universe may not be as detached from the potentialities of life and intelligence as scientists of the twentieth century have tended to suppose.

The Big Bang

Astronomers determine the speed with which a galaxy recedes from us by the redshift of its light. (This frequency lowering is similar to a "Doppler shift," the lowered pitch of the siren of an ambulance that has just passed us. It's actually the expansion of space stretching the light's wavelength.) Astronomers correlate an object's redshift with its distance from us by studying the redshifts of objects whose absolute brightness, and therefore distance from us, is known. They find that the most distant objects we can see, galaxies moving away from us at close to the speed of light, emitted the light we now receive some thirteen billion years ago. Those galaxies were probably about one billion years old at the time. This suggests the Big Bang—the expansion of space that started violently within a small region around fourteen billion years ago.

The strongest evidence for the Big Bang is the cosmic background radiation discovered in 1965. By the time the universe was 400,000 years old, it had cooled enough to allow electrons and protons to combine into neutral atoms. The radiation created in the initial superhot fireball then became independent of matter. At this point the radiation was largely in the visible region of the spectrum. Since that time, space has expanded several thousand fold. The wavelength of that light has been stretched by that factor to become the cosmic microwave background now shining down on us from all directions. The fine details of that microwave background radiation strikingly confirm properties calculated for the Big Bang.

Speculative theories of "inflation" today deal with the immediate aftermath of the Big Bang to explain the remarkable uniformity of the universe on the largest scales. According to these ideas, space almost instantly expanded, or "inflated," at a rate much faster than the speed of light. Starting from something vastly smaller than an atom, the entire universe we observe today inflated in the tiniest fraction of a second to the size of a large grapefruit.

From then on, presently known physics seems able to account for what happened. By the time the universe was one second old, quarks combined to form protons and neutrons. Minutes later the protons and neutrons came together to form the nuclei of the lightest atoms: hydrogen, deuterium (heavy hydrogen, one proton and one neutron), helium, and a bit of lithium. The relative abundance of hydrogen and helium in the oldest stars and gas clouds agrees with what we would expect from this creation process.

Though there is as yet no accepted theory for that minuscule split second before quarks and electrons came into existence, there are constraints on how the universe must have started. To produce a universe resembling one in which we can live, the Big Bang had to be finely tuned. How finely? Theories vary. According to one, if the initial conditions of the universe were chosen randomly, there would only be one chance in 10^{120} (that's one with 120 zeros after it) that the universe would be livable. Cosmologist Roger Penrose has it vastly more unlikely: The *exponent* he suggests is 10^{123}. By any such estimate, the chance that a livable universe like ours would be created is far less than the chance of randomly picking a *particular* single atom out of all the atoms in the universe.

Can you accept odds like that as a coincidence? It would seem more likely that something in yet-unknown physics determines that the universe *had* to start the way it did. Such new physics would likely include a quantum theory of gravity. It may well be the long-sought "theory of everything"—the ToE—uniting Nature's four fundamental forces into a single theory. All physical phenomena (*all* phenomena?) should then be explainable—in principle.

The ToE, like other physics theories, will no doubt be a set of equations. Could a set of equations fully satisfy us? Could a set of equations resolve the quantum enigma without somehow involving the conscious observer? Recall that physics' encounter with consciousness is seen directly in the theory-neutral quantum experiment. It arises conceptually prior to the quantum *theory*. No interpretation of quantum theory or even its deduction from a more general mathematical presentation can resolve what we actually experience in the quantum enigma without involving our conscious decision process.

With another perspective on whether a ToE would explain all we see, Stephen Hawking poses a relevant question:

> Even if there is only one possible unified theory, it is just a set of rules and equations. What is it that breathes fire into the equations and makes a universe for them to describe? The usual approach of science of constructing a mathematical model cannot answer the questions of why there should be a universe for the model to describe. Why does the universe go to all the bother of existing?

Let's step away from Hawking's question and from the one we just posed and look at a host of "coincidences" leading to livable worlds beyond those related to the Big Bang. We might just assume that an eventual ToE will predict all we see—whether or not it can "explain" it. We should perhaps therefore just seek the ToE and be satisfied with it. But critics of that blithe attitude have spoken of an anthropic principle. We start with the more easily accepted version.

The Anthropic Principle

Only the lightest nuclei were created in the Big Bang. The heavier elements—carbon, oxygen, iron, and all the rest—were created inside stars, which formed much later. These elements are released into space whenever a heavy star exhausts its nuclear fuel, violently collapses, and then explodes as a supernova. Later-generation stars and their planets, including our solar system, gather up this debris. We are the remnants of exploded stars—we're stardust.

In addition to the fine-tuning of the Big Bang, another bit of luck seems involved in our stellar creation. Calculations had shown that the making of heavy elements in stars could not get even as far as the carbon nucleus (six protons and six neutrons). Cosmologist Fred Hoyle reasoned that, since carbon was indeed here, there *had* to be a way of making carbon. He realized that a then-unexpected state of the carbon nucleus at a certain very precise energy could allow the stellar production of the elements to continue to carbon, nitrogen, oxygen, and beyond. At Hoyle's suggestion, that crucial nuclear state was looked for and *found*.

There are other coincidences: if the strengths of the electromagnetic and gravitational forces were even slightly different, or if the strength of the weak nuclear force were slightly larger or slightly smaller, the universe would not have been hospitable to life. No known physics compels these things to work out just right.

There are many other coincidences that we don't mention. Does everything so improbably working out so perfectly require explanation? Not necessarily. If it didn't just happen to work out this way, we wouldn't be here to ask that question. Is that statement of explanation enough? Such "backward reasoning," based on the fact that we and our world exist, is called the anthropic principle.

The anthropic principle can imply that our universe welcomes life just by chance. On the other hand, some theorize that a large number of universes, even an infinite number, came into existence, each with its own random initial conditions, even its own laws of physics. Some theories have a grand "multiverse" constantly spawning new universes. The vast majority of these universes likely have a physics that is not life-friendly. Does our existence in a rare, hospitable one need explanation?

Here's an analogy: Consider how improbable *you* are—the improbability of someone with just your unique DNA being conceived. (Millions of your possible siblings were *not* conceived. And now go back a few generations.) With those odds, you're essentially impossible. Does your being here need explanation?

Some urge science to shun the "A-word." The anthropic principle, they claim, has explained nothing, and has even had a negative influence. It should therefore be rejected as "needless clutter in the conceptual repertoire of science." We can understand how it can dampen the drive for deeper searches. But anthropic reasoning can sometimes be fruitful. Consider Hoyle's energy-level prediction for carbon.

Objectors to the anthropic principle, what we now can call the "weak anthropic principle," might be even more averse to the "strong anthropic principle." According to this view, the universe is tailor-made for us. "Tailor-made" implies a tailor, presumably God. That may be something to contemplate. But it should not be an argument for Intelligent Design, as is occasionally suggested. Whoever "breathes fire into the equations," would presumably be omnipotent enough to do it properly at the very beginning and not need to tinker with every step of evolution.

We introduced a different version of the strong anthropic principle in quoting Rees at the start of this chapter: *We* created the universe. Quantum theory has observation creating the properties of microscopic objects. Physicists generally accept that, in principle, quantum theory applies universally. If so, all reality is created by our observation. Going *all* the way, the strong anthropic principle asserts the universe is hospitable to us because we could not create a universe in which we could not exist. While the weak anthropic principle involves a backward-in-time reasoning, this strong anthropic principle involves a strong form of backward-in-time *action*.

Quantum cosmologist John Wheeler back in the 1970s drew an eye looking at evidence of the Big Bang and asked: "Does looking back 'now' give reality to what happened 'then'?" The provocative sketch has not lost impact. At the recent conference honoring Wheeler on his 90th birthday, a keynote speaker introduced his talk with Wheeler's sketch.

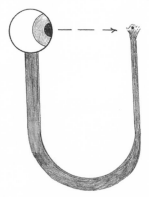

The anthropic implications of his diagram must have been a lot even for Wheeler to buy. After asking the above question, he immediately added the comment: "The eye could as well be a piece of mica. It need not be part of an intelligent being." Of course, that piece of mica supposedly bringing reality to the Big Bang had to be created *after* the Big Bang.

Figure 17.1 Does looking back "now" give reality to what happened "then"?

This strong anthropic principle is in fact too much to comprehend. Though quantum mechanics seems to deny the existence of a physical reality independent of its conscious observation, if our observation creates *everything,* including ourselves, we are dealing with a concept that is logically self-referential—and mind-boggling.

Accepting the boggle, we might ask: Though we could only create a universe in which we could exist, is the one we did create the only one we *could* have created? With a different observation, or different postulate, would the universe be different? It has been wildly speculated that postulating a theory that is not in conflict with any previous observation actually *creates* a new reality.

For example, Hendrick Casimir, motivated by the discovery of the positron after its seemingly unlikely prediction, mused: "Sometimes it almost appears that the theories are not a description of a nearly inaccessible reality, but that so-called reality is a result of the theory." Casimir may also have been motivated by his own prediction, later confirmed, that the vacuum energy in space would cause two macroscopic metal plates to attract each other.

If there's anything to Casimir's speculation, might Einstein's original suggestion of a cosmological constant have *caused* the acceleration of the universe? (Such a speculation can't be *proven* wrong.) Though taking such ideas literally surely seems ridiculous, we see how outrageously the quantum enigma has allowed us to speculate.

John Bell tells us that the new way of seeing things will likely astonish us. It is hard to imagine something truly astonishing that we don't initially rule out as preposterous. Bold speculation may be in order, but so is modesty and caution. A speculation is nothing but a guess until it makes testable and confirmed predictions.

Concluding Thoughts

Quantum theory works perfectly; no prediction of the theory has ever been shown in error. It is the theory basic to all physics, and thus to all science. One-third of our economy depends on products developed with it. For all *practical* purposes, we can be completely satisfied with it.

But if you take quantum theory seriously *beyond* practical purposes, it has baffling implications. It tells us that physics' encounter with consciousness, *demonstrated* for the small, applies to everything. And that "everything" can include the entire universe. Copernicus dethroned humanity from the cosmic center. Does quantum theory suggest that, in some mysterious sense, we *are* a cosmic center?

The encounter of physics with consciousness has troubled physicists since the inception of the theory eight decades ago. Most physicists will dismiss the creation of reality by observation as having no significance beyond the limited domain of the physics of microscopic entities. Others will argue that Nature is telling us something, and we should listen. Our own feelings accord with Schrödinger's: "The urge to find a way out of this impasse ought not to be dampened by the fear of incurring the wise rationalists' mockery."

When experts disagree, you may choose your expert. Since the quantum enigma arises in the simplest quantum experiment, its essence can be fully comprehended with little technical background. Nonexperts can therefore come to their *own* conclusions. We hope yours, like ours, are tentative.

There are more things in heaven and earth, Horatio,

Than are dreamt of in your philosophy.

—Shakespeare, *Hamlet*

Suggested Reading

Blackmore, S. *Consiousness, An Introduction.* New York: Oxford University Press, 2004.

A wide-ranging overview of the modern literature of consciousness from the neural correlates of consciousness, to experimental and theoretical psychology, to paraphenomena. Some mentions of quantum mechanics are included.

Cline, B. L. *Men Who Made a New Physics.* Chicago: University of Chicago Press, 1987.

This light, well-written history of the early development of quantum mechanics emphasizes the biographical and includes many amusing anecdotes. Since it was written in the 1960s, it avoids any significant discussion of the quantum connection with consciousness. (One of the "men" is Marie Curie.)

Davies, P. C. W., and J. R. Brown. *The Ghost in the Atom.* Cambridge: Cambridge University Press, 1993.

The first forty pages give a compact, understandable description of "The Strange World of the Quantum." This is followed by a series of BBC Radio 3 interviews with leading quantum physicists. Their extemporaneous comments are not always readily understandable, but they clearly give the flavor of the mystery they see.

Elitzuir, A., S. Dolev, and N. Kolenda, Eds. *Quo Vadis Quantum Mechanics?* Berlin: Springer, 2005.

A collection of articles, and transcripts of informal discussions, by leading researchers with an emphasis on the paradoxical aspects of quantum mechanics. Some of the papers are highly technical. But aspects of several are quite accessible and indicate how physics has encountered what seems a boundary of the discipline.

Griffiths, D. J. *Introduction to Quantum Mechanics.* Englewood Cliffs, N.J.: Prentice Hall, 1995.

A serious text for a senior-level quantum physics course. The first few pages, however, present interpretation options without mathematics. The EPR paradox, Bell's theorem, and Schrödinger's cat are treated in the afterword. (The book's front cover pictures a live cat; it's dead on the back cover.)

Hawking, S., and L. Mlodinow. *A Briefer History of Time.* New York: Random House, 2005.

A brief, easy-to-read, but authoritative, presentation of cosmology, much of it from a quantum mechanical point of view. Metaphysics and God get substantial mention.

Holbrow, C. H., J. N. Lloyd, and J. C. Amato. *Modern Introductory Physics*. New York: Springer-Verlag, 1999.

An excellent introductory physics text with a truly modern perspective, including the topics of relativity and quantum mechanics.

Miller, K. R. *Finding Darwin's God*. New York: HarperCollins, 1999.

A convincing refutation of Intelligent Design that also argues that arrogant claims of some modern scientists that science has disproven the existence of God has promoted antipathy to evolution, both Darwinian and cosmological. Quantum mechanics plays a prominent role in Miller's treatment.

Park, R. L. *Voodoo Science: The Road from Foolishness to Fraud*. New York: Oxford University Press, 2000.

A brief, cleverly written exposure of a wide range of purveyors of pseudoscience who exploit the respect people have for science by claiming that science gives credence to their particular nonsense.

Schrödinger, E. *What Is Life? & Mind and Matter*. London: Cambridge University Press, 1967.

An older, but very influential collection of essays by a founder of quantum theory, including one entitled "The Physical Basis of Consciousness."

Index